PRAISE FOR *LIFE LESSONS FROM A PARASITE*

"In nature, as in society, the parasites outnumber the hosts. John Janovy Jr. offers the parasites' view of this situation. The result is smart, funny, and all too revealing."

—Elizabeth Kolbert, staff writer for the *New Yorker* and *New York Times* bestselling author of *The Sixth Extinction*

"This book, a love song to research scientists, is written with the hope that we ordinary humans can learn from their work. Janovy implores his readers to base their conclusions on evidence, rather than being swayed by the many infectious and false ideas in our culture. He stresses that successful organisms adapt to, rather than deny, a changing reality. And he stresses the uniqueness of every organism, no matter how small. Most of all he shows us the complexity and beauty of parasites. After reading this book, I'll never look at mosquitoes in the same way again."

—Dr. Mary Pipher, author of *Letters to a Young Therapist*

"Professor John Janovy Jr. is one of the one of the great thinkers in the field of parasitology, and our understanding of host parasite relationships. I'm thrilled that now he has synthesized his thinking in this important new book. It is especially timely given how students in parasitology lately go right to the molecules or the cells without the context of the whole organism biology. His new book will help remind us of what we are missing!"

—Peter Hotez, MD, PhD, DSc (hon), FASTMH, FAAP, author of *The Deadly Rise of Anti-Science*

ALSO BY JOHN JANOVY JR.

Nonfiction

Keith County Journal

Back in Keith County

Fields of Friendly Strife

On Becoming a Biologist

Ten Minute Ecologist

Vermilion Sea

Dunwoody Pond

Comes the Millennium (as Jack Blake)

Intelligent Designer: Evolution for Politicians

Foundations of Parasitology (with Larry Roberts and Steve Nadler)

Teaching in Eden

Outwitting College Professors

Pieces of the Plains

Africa Notes

Letter to a Child Born Today

Fiction

Yellowlegs

The Ginkgo

The Gideon Marshall Mystery Series

Conversations Between God and Satan

Tuskers

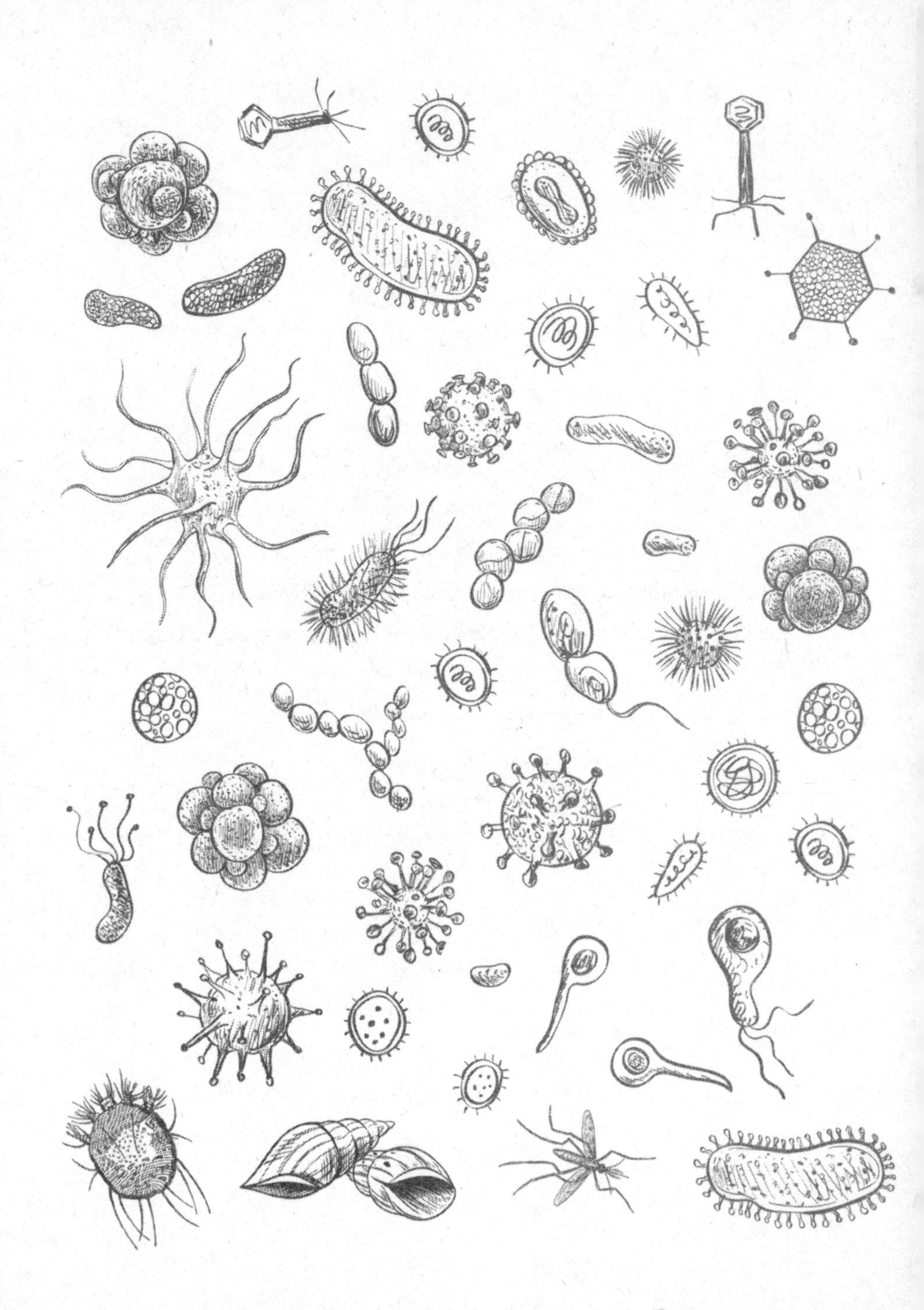

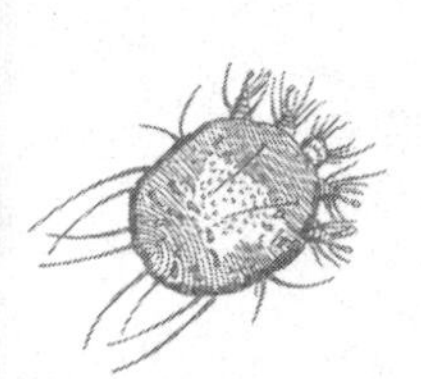

What Tapeworms, Flukes, Lice, and Roundworms Can Teach Us About Humanity's Most Difficult Problems

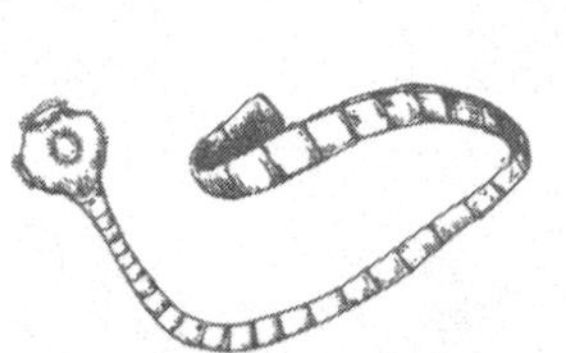

LIFE LESSONS *from a* PARASITE

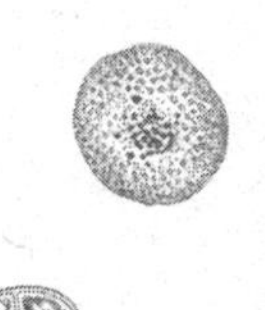
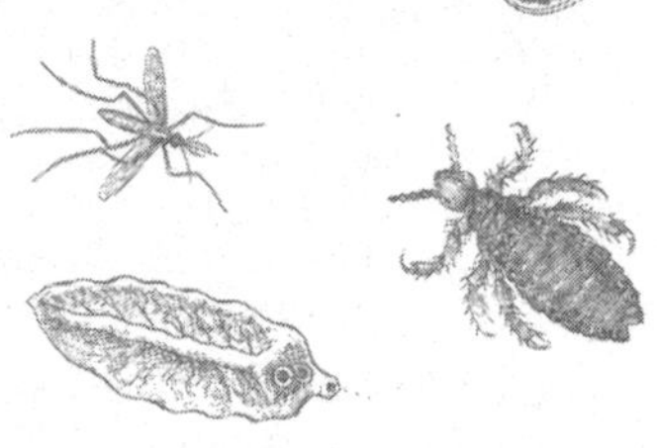
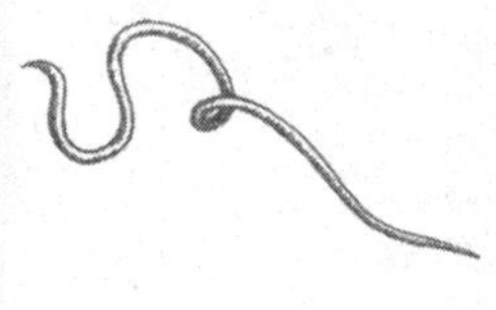

JOHN JANOVY JR. PhD

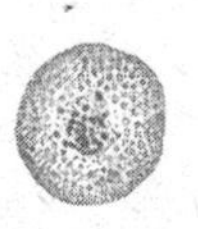

sourcebooks

Cover design by *the*BookDesigners
Cover images © aksol/Shutterstock, Bodor Tivadar/Shutterstock,
Hein Nouwens/Shutterstock, Morphart Creation/Shutterstock,
Prokhorovich/Shutterstock, Robusta/Shutterstock
Internal design by Tara Jaggers/Sourcebooks

Published by Sourcebooks
P.O. Box 4410, Naperville, Illinois 60567-4410
(630) 961-3900
sourcebooks.com

Cataloging-in-Publication Data is on file with the Library of Congress.

Printed and bound in the United States of America.
SB 10 9 8 7 6 5 4 3 2 1

This book is dedicated to the two hundred or more students, from first-year undergrads to doctoral candidates, who worked in my lab from 1966 to 2015 delivering all those life lessons about science, college, and parasites just by following their interests and curiosity.

CONTENTS

INTRODUCTION

A World of Infection

Your oldest child comes home from school with head lice, along with a note that they can't return until the lice are gone. After the screaming subsides, your trip to Google turns up some options, including shampoos, combing, Vaseline, mayonnaise, and an insecticide—pyrethrin, derived from chrysanthemum flowers, or permethrin, the synthetic version of it—recommended by the Centers for Disease Control and Prevention.[1] But then there's the problem of lice eggs, or nits, still stuck to hairs and pillowcases.[2] If you're a fan of ancient history, you now have an opportunity to comb through your child's hair and go nit-picking, just like early humans did and chimpanzees still do.[3] Regardless of how you handle this situation, you'll come away educated about parasites—species that must live in or on other, unrelated species to survive—and parasitism, the most common way of life among animals on Earth. That education, delivered by lice, should be useful when your other child comes home from day care with pinworms or from high school infected with bad ideas.

Parasites are organisms that live in close, even intimate and obligate, relationships with unrelated species—their hosts. That relationship is both physical and physiological, involving parasite use of host tissues and substances such as sugars and amino acids. Furthermore, parasites cannot survive without this relationship. Why do we believe that parasitism is the most common way of life among animals on Earth? The answer is simple: every species that has been studied carefully has been shown to be infected with several different kinds of worms, lice, ticks, mites, and single-celled organisms known as protists, all in addition to bacteria, fungi, and viruses. For example, at least forty different kinds of parasites—tapeworms, roundworms, lice, etc.—have been reported living in and on the robins that show up in North American backyards.[4] Similar claims can be made for all familiar birds, but even the insects in your garden and earthworms down in the soil have parasites. How do those animals get infected? That's the same question you'll ask about your children with head lice and pinworms. That's the same question you'll ask when your children are older and get infected with ideas that change their behavior and are not consistent with your dreams for their futures. That's the same question the world has been asking while a virus sweeps across continents leaving a trail of ignorance and death in its wake.

This book is also about scientists who spend their lives trying to answer questions of transmission, in the process learning important lessons from the parasites they study, lessons applicable far beyond the field of parasitology. These scientists are captivated by what they find when they cut up some animal and start digging through its guts. Although their immediate interests usually concern how these animal parasites are sustained in nature, it doesn't take much of this research before a person starts making a connection between parasitic worms and parasitic ideas. So this book is also about what the study of parasitism does to scientists' minds, especially their thoughts about all kinds of infective cultural phenomena, including memes, fake news, conspiracy theories, and weapons, to name a few.

Anyone who's studied parasites of any kind sees a common thread among things that move through populations: infectivity is a property that is possessed or can be acquired by almost anything we humans can discover or make, including our pictures, stories, music, machines, and words, all in addition to those viruses, bacteria, fungi, and nonhuman animals that can and so often do make our bodies their homes. This book is my attempt to lead you to the same conclusion not only by letting you share the experiences of some chosen parasitologists but also by proposing a general theory of infectivity that applies to whatever we encounter during our lives.

But to really understand this ubiquitous way of life, one needs to collect animals, look through their bodies for the other animals living there, and pay attention to what those parasites are telling us about how they live and prosper. Unfortunately for those host animals we're taking lessons from, they usually must be captured and killed, sometimes in large numbers, before we can draw conclusions about the lives of infectious agents, especially in populations. Once during a bar conversation over beers with some young scientists, we tried to estimate the numbers of wild animals that we'd collected while searching for parasites. At the time, I was a senior citizen, but my colleagues were in their late twenties or early thirties, pursuing advanced degrees in biology. Although their time for exploring the natural world was less than mine, a couple of them had done far more collecting, and subsequently dissecting, than I could account for personally because they were studying nature in a manner that required hundreds of specimens to satisfy statistical requirements. What were these studies? They were ones that focused on the way these infectious organisms developed and were transmitted from one host animal to another, spreading through an ecosystem.

Whether they realized it or not, these young scientists were preparing themselves to understand global pandemics, evaluate political response to epidemic disease, and serve as teachers dedicated to shutting down a pathogen's opportunities for infection. In the process, they'd come to view living organisms not as individuals or even species but as communities. If you're a parasitologist on a downtown street corner, for example, and across the street are others waiting for a green light to cross safely, what do you see? You see members of our human species, a few examples of the nearly eight billion other people who now occupy Planet Earth and collectively provide a nutritious environment for at least two hundred and fifty different, yes *different*, kinds of other animals—parasitic worms, lice, mites, and their ilk, all in addition to viruses, bacteria, fungi, and got-it-from-my-friend ideas about treatment of diseases like COVID-19.[5]

Are these parasites thriving in the people you now see walking toward you? Maybe, but probably not if you're in someplace like Omaha, Nebraska. Because of their location and good fortune, these folks are walking examples of the first rule of infectious disease transmission: standard of living matters. Functional sewage systems, window screens to keep out the mosquitoes, shoes to stop hookworm juveniles from burrowing into the skin between your toes, water treatment plants that eliminate cysts and spores, and affordable health care are all manifestations of a relatively high standard of living. But standard of living also includes an educational system that provides skills needed to sort valid information from outright lies, outlandish claims, and conspiracy theories that populate social media and move through nations as easily as a pandemic virus. This book is my attempt to add that last skill to your standard of living.

Most of the material in the following chapters is derived from my research activities as a biological scientist over a fifty-year period, which in turn rested on a childhood spent studying nature as a hunter, trapper, fisherman, and addicted reader of hunting and fishing magazines, books about firearms, and adventure fiction such as the now politically incorrect *Bomba the Jungle Boy* series. That behavior was aided and abetted by a father and paternal grandfather, the former with gear and what seems in retrospect to have been an obligation to teach his son to hunt and fish, and the latter with stories of wild early twentieth-century life on the American Great Plains. Conversation in our home was mostly about literature, nature, geology, music, and global affairs. Thus, looking back on the sixty years since I made the decision to be a professional biologist, that Oklahoma childhood played out on the edge of town, often with a fishing pole or shotgun in hand, seems to have been a preview of what I would learn about animals as a scientist. *Life Lessons from a Parasite* is my attempt to describe that revelation and arrive at an attempt to answer the question of what it means to be a human with power to wreak havoc on other living organisms, including our fellow humans.

One of the first lessons I learned from studying parasites is that our personal reaction to collecting wild animals—an act that enables us to look for worms within the victims' bodies—depends on a number of factors, including who or what is being killed, where it is being killed, why it is being killed, how it is being killed, and who is doing the collecting. In writing this book, I've tried to address these questions of who, what, why, when, where, and how and to address them in a way that leads naturally to the question of what we, as human beings, do to build and sustain an individual, national, ethnic, or religious identity. I contend that giving some serious thought to why, when, where, and how I might disassemble a beetle or a small fish leads inevitably toward serious thoughts about why my fellow humans behave as they do, especially with regard to the taking of lives. Something happens to generate this comparison, something worth writing about, talking about, and thinking about daily. That something is part of what makes us *Homo sapiens*, and it's also why I'm writing this book about the collection of animals and the rationale behind such collecting, namely in an attempt to satisfy our curiosity about parasites and parasitism, eventually including infection with dangerous ideas.

The first few chapters include stories of people studying the lives of parasites, that is, being parasite hunters, mostly in the North American Great Plains. I've tried to bring you into their worlds and into mine as we collected animals and extracted the other kinds of animals living in and on the ones we'd collected. I've also tried to engage you in the driving curiosity about parasitism that in turn drives parasite hunters' actions. Throughout those chapters, I've tried to lead you intellectually into the kinds of thoughts and conversations this research produces in its practitioners—thoughts and conversations about larger questions, such as why we are so vulnerable to infection with dangerous ideas, especially ones that spawn hate and violence.

In later chapters, I've tried to elaborate on aspects of science that are rarely, if ever, dealt with in either popular or scientific literature, namely that realm of work never published but that nevertheless has a powerful impact on the thinking of those involved in doing it or watching it being done—thinking that leads naturally to long discussions about the nature of science and especially the influence that investigation of the natural world has on the people doing it. Parasitology is a rich, challenging, and often inaccessible discipline practiced by those who are not afraid of its mysteries and potential failures. As a University of Toronto biochemist once remarked in a speech honoring the retirement of a colleague, a distinguished and well-known parasitologist: "Parasitology! Queen of biology! One part of science to two of mythology."[6] Those of us who practice it understand.

Thus, those later chapters are my attempt to engage you in the kind of mental extensions that come from a lifetime spent studying infectious organisms and living professionally with others trying to do the same. Very few of the parasites introduced in the following pages are infectious to humans or our domestic animals. I've never been a scientist for the self-proclaimed purpose of curing humanity of its burdens, so you're not going to read very much about deadly disease or terror in your drinking water. Instead, all this collecting, dissecting, and subsequent lab work rests on the well-known fact that parasitism is the most common way of life among animals on Earth and on curiosity about how those parasites survive in nature. If parasitism is the most common way of life, goes the reasoning, it should be found everywhere, in abundance and diversity, and its various ways of managing existence, along with the accompanying lessons, should lead us naturally to questions about ourselves.

JOHN JANOVY JR.

CHAPTER ONE

WHY PARASITES?

Nature Ignores Your Desires

Find the parasite first, and everything else will fall in place.

Gerald W. Esch, *Parasites, People, and Places: Essays on Field Parasitology*, 2004[1]

Parasitology is the study of how and why an organism's body can become occupied by another, unrelated organism. Parasitologists tend to focus on this relationship between host and parasite. We try to determine how that relationship is maintained, how a parasite gains access to a host, and whether the host has defense mechanisms to help protect it from invasion. We also ask the evolutionary questions of why this relationship was formed in the ancient past and what host species are now occupied by a parasite's relatives. By trying to discover not only how and why one species comes to live in or on the body of another but also how these parasites move through populations of hosts, we learn lessons that are not confined to the host and parasite. We can apply these lessons to animal life in nature, human life in society, ourselves, and problems far more serious than, for example, a mysterious worm inside a fish in an Oklahoma lake.

Complex life histories are the main reason why parasitology is such an educational discipline. Many parasites have what are called indirect life cycles because to develop into an infective stage, they must spend time inside more than one host. A parasitic worm's eggs can hatch into larvae that develop only to a certain point in one host species, which then must be eaten by another host species before the worm can develop further to become infective for the next host, and so on. To understand how such parasites are maintained in nature, we must study all these life cycle stages, including their ecology and evolutionary history. This "homework" turns parasitologists into broadly educated people. It also, however, tends to guide their research by making them expect certain events to happen in nature and giving them hope they can replicate those events in the lab.

Scientists who study parasite transmission also come to view all things that move through populations as infective agents carried by hosts. Hands-on experience with worms that live inside other animals and have complex life cycles teaches us to study every possible avenue for parasite transmission. Eventually we learn that infected hosts can carry everything from viruses to fleas to lice and worms, as well as ideas that alter human behavior. For this reason, parasitologists are altered by their research, but that fate of being shaped by the work you do is not exclusive to parasitologists or scientists in general or to the rest of us regardless of our chosen trades or professions. Artists are created in large part by what they learn from their art, musicians from the lessons their music teaches them, and athletes from the way their games are played. But only parasitologists spend their days studying the most common way of life among animals on Earth, so it's not surprising that they end up with a diverse repertoire of life lessons, including ones taught by worms that defy scientists' best efforts to explain their existence.

The following pages tell a story in which these expectations were never met, with constant failure delivering an important lesson about nature and about ourselves. The scientists involved finally had to admit that what seemed like a simple problem in the beginning turned out to be a monumental mystery that no amount of effort was able to solve.[2] This problem is a seemingly simple one: A species known as buffalo fish lives in an Oklahoma lake and has strange worms in its ovaries. A scientist wants to discover how those worms got there. He doesn't want to cure the buffalo fish of this infection or make you believe he can do that. He's just captivated by the odd structure of these worms and their unusual location—a fish's ovaries. The worm is not infective to humans, so it poses no health problem to people who might eat the fish or inadvertently gulp down the water while swimming in the lake. The problem worm has a scientific name, *Nematobothrium texomensis*, which, like any of the hot-button words that show up in daily headlines (such as "immigration" and "climate change"), hides a complex problem. Yet the issues defined by these words end up being portrayed in simple terms, especially by elected officials with agendas. These words are also examples of infective agents but ones with all the personal, social, and economic conditions they contain hidden in and carried by language instead of in a fish that has parasites packed away in its ovaries.

This worm story takes place mostly in southern Oklahoma. Denison Dam connects Oklahoma and Texas, from north to south, respectively, and blocks the appropriately named Red River, producing one of the world's most ramified freshwater bodies, Lake Texoma. Home of this parasitic worm that so far has defied everyone's efforts to explain its existence, Lake Texoma spreads into the northern Texas and southern Oklahoma valleys west of Denison Dam, providing a recreational bonanza. Its hundred and forty square miles of open water are surrounded by places like the Chickasaw Point Golf Course, the Roads End Public Use Area, and the Hook, Wine, and Sinker bar.[3] Beneath that open water surface, which is often wind-whipped to a fury that keeps sailors hunkered down over their beer, swim buffalo fish, so named because their profile reminds us of the iconic American bison. And within those buffalo fish lives this strange worm, so structurally different from its relatives, occupying the fish's ovaries instead of an intestine and disintegrating to release its eggs instead of just releasing them from a uterus.

The scientists who tried but failed to solve the mystery of its existence—how it gets from fish to fish—are all dead, but the worm remains. Thus, we have the first of many life lessons taught by parasites: *Seemingly simple problems can have solutions that elude the finest minds.* That this lesson involved a parasitic worm in a fish makes it seem almost inconsequential compared to the other problems scattered across our daily news feed. But just like the rest of us have failed to solve those larger and much more consequential problems, the scientists obsessed with this worm failed to discover how it was transmitted through the fish population that allowed it to survive in the depths of Lake Texoma. Their years spent trying, however, using every means at their disposal, taught them this lesson in a way they'd never have truly appreciated just from reading about those larger problems.

This parasitic worm resembles a nearly microscopic thread in the buffalo fish's ovaries.[4] The fish routinely spawn and reproduce; there is no evidence that the worm damages its host. The scientific literature—playing an authority role—seems to suggest how this parasite lived in nature, how it developed and became infective to the next fish. So the scientists trying to solve the problem of transmission trusted the literature and used it as a guide to design their studies. That literature led them astray, not unlike what has happened before in medical research. From such leading comes another lesson taught by a parasite and applicable, perhaps, to the words "climate change": *What happens in the natural world has no necessary connection to your beliefs, your desires, or what others have said.* That lesson applies even when we contribute to the natural world's condition, for example, by burning fossil fuels or by destroying tropical forests at a hellish rate.

This recalcitrant worm is a fluke that does not match our perception of either its structure or its life in the wild. The parasites we call flukes are flatworms; they generally range in size from near microscopic to an inch or so in length, and their flattened shape makes many of them resemble leaves. Adult flukes typically have both male and female reproductive systems in a single individual. Cross-fertilization is evidently an advantage when your eggs' embryonic development depends on encountering and then occupying a sequence of specific living environments—mollusks, crustaceans, aquatic insects, other fish that serve as intermediate hosts—and when the chances of such encounters are near zero for any single egg. Parasitic worms are notorious reproductive machines, spewing vast numbers of eggs into the water or on land, depending on where their host lives, usually by way of the host animal's feces.[5] The Lake Texoma worm differed from expectations, however, by releasing its eggs through a female fish's ovarian ducts, and only at spawning time. That difference should have been a warning to the scientists investigating it: *Expectations can hinder the search for the truth.*

In the summer of 1961, Dr. John Teague Self, known to his

professional associates as "Teague" and to everyone else as "Dr. Self," was a bear of a man, both physically and mentally. When ensconced behind his desk on the second floor of Richards Hall at the University of Oklahoma, he lowered his head and stared at you evenly over his glasses if you had the courage to come to his office. At social gatherings, we students would mimic his trademark over-the-glasses stare. Among the graduate students who gathered at Across the Street, one of the finer college town bars in Norman, Oklahoma, to review their week's successes and failures, someone would inevitably mention this buffalo fish worm, which Self had discovered and whose existence defied explanation because he could not discover its method of transmission. The Across the Street crowd included those who had not been sucked into Self's obsession with *N. texomensis*. So they were happily drinking beer on a Friday afternoon instead of setting nets in Lake Texoma. Yet they still wondered how this parasite could have such a powerful effect on a person's mind, his curiosity driving him even in the face of failure.

That power may have been derived from the fact that everything about *N. texomensis* violated the textbook picture of flukes. This worm was twelve feet long, thin as a pencil lead, and lived in the ovaries of Lake Texoma buffalo fish. It matured only at certain times of the year, when the fish were spawning and their ovaries were filled with eggs. As the fish shed its eggs, the worm died and disintegrated, and its pieces, containing infective worm eggs, were spread out into the lake. All this worm's relatives live in ocean fish, but *N. texomensis* occupies a freshwater fish with no direct link to the ocean, its habitat in the ovaries is unusual, its disintegration as a means of shedding eggs is unique, and its thread-like structure is uncharacteristic of flukes.[6] The mystery fueled Self's obsession. The worm drove him to do years of research, infecting his mind with the idea that with enough effort, he could solve this problem. But among scientists who study parasites, that's a fairly common occurrence.

The *N. texomensis* story is told officially in a few papers published in the scientific literature, including the *Journal of Parasitology*, but the unofficial story is written in a small field diary from the early 1960s, filled with Self's characteristic cursive.[7] I've borrowed this field notebook from the archives of the University of Nebraska State Museum, which houses a research collection of parasites of global importance. The curator, Dr. Scott Gardner, refers to Self's diary as "the codex" and writes out a legal notice confirming that I have taken this book from fire-resistant metal cabinets that house other irreplaceable treasures. That notice is a warning: treat this field notebook as if it is an original copy of the Declaration of Independence. Like the authors of that declaration, most if not all the people who penned entries into Self's codex are dead. The others, who I believed might still be living and tried to contact, never responded; eventually I decided they were embarrassed at having been captured by the idea of this worm and refused to admit it.

But there are long sections of handwriting other than Self's in this small brown K&E surveyor's field book, and after reading through some of it, I discover the initials "C.E.D." for Clarence Edgar Davis, a research assistant on the *N. texomensis* project, at the time funded by the National Science Foundation, that is, with government money. Davis's writing in this notebook is that of a mature scientist—careful, detailed, legible, in permanent ink and perfect grammar, and initialed "C.E.D." after each entry. I can almost hear him thinking as he makes those entries, probably at night after a long day of collecting and dissecting fish. He knows someone else will be reading his notes; he doesn't know a person will be holding that little book half a century later and wondering how in the living hell anyone as smart as all these people, including the anonymous reviewers who delivered that money for Self to spend on this project, could get caught up in the life of this seemingly unimportant worm that causes no damage to humans. But in 1960, *N. texomensis* was of serious importance to at least two humans: J. Teague Self and Clarence Edgar Davis.

At the time I knew him, I believed that Clarence Davis was Self's graduate advisee as well as research assistant on the *N. texomensis* project and that he must have been on leave from his faculty position at Southern University in Baton Rouge, Louisiana, a historically Black institution, to pursue his doctorate. Davis's work required that he drive a state-owned vehicle alone to Lake Texoma, set gill nets, collect buffalo fish, handle them in a specified way, and otherwise do the work of a young scientist. Only in retrospect do I have a sense of what kind of courage and social skill were required for him to do what Self expected him to do, given that the setting was rural southern Oklahoma in the late 1950s and early 1960s. I remember standing in the main Oklahoma City library stacks on May 17, 1954, reading a book, when a perfect stranger came up and started shaking my arm, telling me in vulgar racist language how different my life would be now that the Supreme Court had let little Linda Brown into the elementary school closest to where she lived. That's the cultural landscape that Davis had to negotiate to do his research, do his job, and satisfy the curiosity he had about the life of worms in fish.

Nowadays, the national news is a constant reminder that the challenges Davis faced trying to do his scientific work in that 1950s cultural landscape have not disappeared. During the time this book was written, media reports of racially charged language by elected officials were a regular occurrence, as were voter-suppression tactics aimed at reducing participation of minorities in our nation's democratic processes. The current proliferation of weapons and the readiness with which some of us use them to counter threats such as a child asking for directions suggests that Davis would be putting his life in danger every time he got into a state-owned vehicle alone and headed south out of Norman, away from the relative safety of a college campus, just to collect fish and look for worms in the ovaries.

Had I known then what I know now about the evolution of my nation, I'd have asked Davis if I could ride along on one of his trips to set gill nets, curtains of death that hang down from a float line into the murky waters of Lake Texoma. Fish that swim into these nets get trapped because once their front end goes through one of the square openings, their gill covers expand, get caught on the fine filaments from which these nets are made, and prevent the fish from backing out. Setting those nets and retrieving them is major physical labor done from an aluminum boat made for two people. The goal is to have a dozen or so large fish upon retrieval, and each fish represents at least an hour, maybe two, of processing. But at the time, I believed Clarence Davis was just another grad student; in my mind, he was "working on *N. texomensis*." Our projects become our identities; that fact has not changed in the decades since I met Self.

My own use of gill nets helps me imagine the scientific part of Davis's travels half a century later. On a fall afternoon, he leaves the University of Oklahoma Biological Station (UOBS) building, crosses a gravel drive, and walks down a path to the UOBS boathouse. He fires up a boat engine and breathes in the exhaust that reminds him of where he is, what he'll be doing for the next couple of hours, and most importantly why he's doing it. Davis smiles, alone on the water, knowing that along with Self, he's caught in the intellectual net that *N. texomensis* has spread, waiting for curious humans to swim up and encounter the mystery. Out in the lake arm known as Buncombe Creek, he cuts the engine to idle and reaches into the tub where his net lies perfectly coiled from the last time he used it. Half a concrete block is tied to a bottom line. He drops the block into Lake Texoma, then picks up the float made from a plastic jug and drops it over the side too. He shifts into reverse and backs up slowly, letting the net play out, its top held up by floats, its bottom sinking down eight feet, pulled by a leaded line, spreading this curtain of death a hundred yards across Lake Texoma.

While I can envision his scientific activity, I have no idea what Davis does for dinner or what happens when he tries to buy gasoline for his state-owned vehicle. The nearest town is Kingston, Oklahoma; even had I been with him, he may not have been able to go into a café with me there or get served if he did. Maybe he's brought along groceries that don't need to be cooked—peanut butter and bread, Vienna sausages, and a jug of water from Norman, where water doesn't taste like southern Oklahoma gypsum. Nothing in my experience lets me envision what he does when he's not catching and dissecting fish. The only thing I can share with Clarence Davis is the reason he's chasing this worm: curiosity drives us, and nature provides an opportunity to study the object of that curiosity and convert our observations into scientific facts.

Out on the lake the next morning, Davis slowly pulls in the net, removing the fish and tossing some of them into a bucket. He'll dissect the four bigmouth buffalo he's caught—*Ictiobus cyprinellus*—for worms. He saves the four-pound channel cat; he'll clean it, put the meat in a lab freezer, and live off it for a week back in Norman.

A couple of crappies get tossed back into Lake Texoma; one of the crappies floats, dead, collateral damage and snapping turtle food.

Back in the dorm building, he goes into a lab and starts on his first fish. It's still alive, wet, slippery with mucus, and flipping, two and a half pounds of live animal. He picks it up with both hands, his left thumb in the fish's mouth while his left-hand fingers make a fist under the throat. His right hand picks up a pair of heavy, stainless-steel scissors with short, stocky, blades and long handles to provide leverage and slips one of the blades into the upper gill chamber before pushing with all his might, crushing the backbone behind the head. That cracking sound is a familiar one that he remembers from the first time he dissected one of these fish, and it's the quickest, thus most merciful, way to kill a large fish. The body still flips; fish nervous systems are not as centralized as those of mice.

After a few seconds, the movement stops; Davis pushes one scissors blade into the anus and starts cutting forward, up to the gill chamber. He spreads the cut open, revealing all the internal organs, an anatomy lesson that he's learned well. Gravid female; ovaries bulging with eggs. He clips off the ovaries and moves them to separate petri dishes. Whatever else the fish may have been, whatever role she may have played in Lake Texoma, is erased; this female buffalo fish is now two ovaries in petri dishes.

Now Clarence Davis is asking the same questions everyone who has ever studied parasites has asked: are there worms in there, and if so, how did they get here? Those questions are dangerous; they can produce an all-consuming struggle with some wild mystery that Mother Nature divulges only grudgingly, if at all, as Self's experience demonstrates. But those questions are the same ones raised by any problem with infectious agents, be they worms or memes; only the methods for answering those questions differ, and the people who study the latter are more likely to be historians or marketing gurus than parasitologists.

How does something get somewhere you'd never expect it to be? How does someone achieve success against all odds and against your expectations, given their origins, constraints, and opportunities? Although research on worms such as *N. texomensis* can easily be ridiculed—"Why the hell are they wasting their time and my tax money on *that*?"—the simple truth is that such research accomplishes two things of importance to a nation heavily dependent on science and technology. First, it demonstrates that a scientist can do the fundamental behavior of science successfully—develop testable hypotheses, assemble the resources necessary to test them, and bring a project to closure with publication after peer review. In the opening years of the Common Era's third millennium, the United States is in desperate need of a massive supply of such human resources and the kind of rationale they bring to the dialogue about our nation's problems, especially during a viral pandemic. Davis's encounter with *N. texomensis* functioned in that expected manner, providing him with insight that would be passed on to his students for the rest of his career.

The second thing academic research accomplishes is that it keeps faculty brains in shape. Faculty brains are the most valuable and costly resources of American higher education, the enterprise that produces the nation's engineers, physicians and other healthcare professionals, attorneys, public school teachers, and accountants, to name just a few. If you were ever inclined to assign power to a worm, *N. texomensis* would be a candidate for champion scientist trainer, the intellectual equivalent of a multimillion-dollar football coach.

The main difference between the two is that if the coach wins enough games, he gets a big raise, but when a worm wins, that only deepens its allure, in essence infecting a scientist with an obsession.

In their first published report about the worm, Self and a colleague, Allen McIntosh, a federal Department of Agriculture scientist, described it as so delicate that it could not be recovered in one piece. By putting two specimens' pieces together, the scientists concluded that they were about eight feet long and one millimeter (four one-hundredths of an inch) wide. They could distinguish internal anatomy features, which then allowed them to figure out what general group of worms it belonged to—the flukes, known officially as Trematoda. They were also the first ever to describe it in the scientific literature.[8]

This physical description was striking for a fluke. Imagine that you had personally handled three thousand different kinds of flukes from a thousand different kinds of animals, and all of them were flat, no more than an inch long, and mostly a quarter to a half inch wide. Now, from a buffalo fish, here is a worm eight feet long and only as thick as a piece of string. And instead of living in a host's intestine and constantly spewing eggs by the thousands like most flukes, this one lives in a host's ovaries and releases its eggs only when it dies and breaks apart and the fish spawns, shedding its ovarian contents. Everything about this worm violated expectations, compounding the mystery of its existence.

Although Self and McIntosh published the original species description, much of the difficult fish collection and dissection work was performed by Clarence Davis. Bigmouth buffalo fish can live for a hundred years, reach four feet in length, and weigh eighty pounds, although the ones from Lake Texoma were generally about two feet long and four or five pounds. Dissection of a single infected fish could easily take an hour because of the worm's delicacy and the care required to extract it from the ovaries. On one trip, Davis worked with a commercial fisherman to set nearly a mile of net. The next day, he dissected thirty-two females and thirty-six males. Two of the females were infected; none of the males were, but he had occasionally seen worms associated with the testes. On other trips throughout the year, he processed anywhere from forty to sixty fish, reporting that the worms grew slowly through the winter, maturing and mating in the spring and then disintegrating into pieces, with their presumably infective eggs, when their host spawns. Fish eggs and worm eggs get flushed out into Lake Texoma together; development of a parasite and host are linked physiologically. We know this fact—a lesson on how the world operates—because of Davis's work.[9] What we don't know yet is how those worms get into buffalo fish or why they are not found in all the other species of fish in Lake Texoma.

"How" questions are ones involving mechanisms, the way things work, and in general they are relatively easy to answer unless, of course, it's the one involving the life history of *N. texomensis*. "Why" questions are the serious ones, questions of origin, evolutionary questions. Most scientists concern themselves with "how"; the answer to "why" can become as all-consuming, elusive, and ethereal as *N. texomensis* in a fish's ovaries. In the case of trematodes—those parasitic flatworms we call flukes—"how" typically involves snails or clams that become infected with larval worms that mature to a certain stage before being released. In some groups, additional hosts such as insects or small fish must acquire these larval worms and then be eaten by a final host before the parasite can become fully mature, mating and spewing hundreds or even thousands of eggs into the environment.

"Why" questions involve evolutionary histories and long time periods during which potential parasites become adapted to live in or on particular hosts or at sites within a host species' body or acquire avenues of transmission. These adaptations involve genetically embedded traits that cannot always be recovered by molecular methods yet can determine not only which host species can and do contain a particular kind of parasite but also how those parasites are transmitted. Not all flatworms are parasitic; planarians, commonly used in college biology class animal behavior lab exercises, are free-living examples. The origin of parasitism in flatworms is an evolutionary mystery, but it probably started with worm-snail associations that eventually involved invasion of snail bodies by worms that then proliferated by asexual reproduction. The result, many millions of years later, is that the vast majority of trematodes must reproduce inside snails before continuing their life cycles in a wide variety of insects, crustaceans, fish, and other vertebrates.

Snails are also notorious as vectors for parasites that cause some important diseases of humans and domestic livestock. Good examples of such worms are human blood flukes known as schistosomes and liver flukes in cattle, sheep, and goats. Human schistosomes infect well over two hundred million people globally, destroying their livers and stunting their growth, with up to seven hundred million at risk in seventy-eight countries. Because larval schistosomes leave their snail intermediate hosts and penetrate skin directly, cultural practices such as rice cultivation contribute to transmission. Domestic animals acquire liver flukes by eating vegetation upon which the parasites have encysted after leaving the snail. Losses due to veterinary infections are in the millions of dollars globally.[10] So regardless of how bizarre *N. texomensis* was structurally and how different it was from other flukes, Self and his students started trying to solve the life cycle problem by collecting snails.

By starting with snails, these scientists were thus behaving like so many of us do when faced with a problem: try a solution based on what we've already learned from a variety of sources, including parents, friends, school, and the media. In this case, the idea that snails would be a first intermediate host for flukes is a pervasive one, taught in every parasitology course and transmitted through sources with apparent authority, such as textbooks, lectures, and published scientific literature. J. Teague Self didn't need to get infected with an expectation that snails would be the first intermediate hosts for some trematode species; he'd transmitted that idea himself, through his teaching, and had seen the reality of snail-trematode interactions in nature through his own research.

Trematode eggs shed into water are either eaten by a snail, after which they hatch, burrow through the snail's intestine, and start reproducing in the digestive gland—the snail's equivalent of a liver—or hatch into a larval form that then penetrates the snail before reproducing in the digestive gland. Later larval stages leave the snail and enter another host, sometimes by penetration, sometimes by being eaten, and sometimes by being eaten after having already penetrated another host. Human lung flukes are a good example of this sequence because people get infected by eating poorly cooked crabs that in turn have been infected with larvae from snails.[11] If human lives seem fraught with risk, trematode lives put that chanciness into perspective, although in a rich and diverse aquatic ecosystem, there are usually enough snails and other small invertebrates to ensure transmission of the species, regardless of how little chance there is that a single worm egg will ever develop into an adult.

Self and the others, including both Davis and another grad student named Lewis Peters who had been caught up in this intellectual trap, focused their work on what they knew about flukes in general, namely that life cycle pattern described earlier. They started snail experiments to discover if what should happen actually did—and of course it did not. Eggs from *N. texomensis* were plentiful, provided someone collected female fish at certain times of the year. Snails of several different kinds were also plentiful, easily collected, and cooperative enough to stay alive in the lab, sometimes even reproducing. So the lab began the experimental infections. Worm eggs were fed to snails; no infections resulted. Worm eggs were put into water with snails; none hatched into larvae that burrowed into the snails. Someone said what was obvious to all: maybe we don't have the right kind of snails. Self, Peters, and Davis became snail farmers, collecting and raising every species they could collect from Lake Texoma, the streams that fed into the lake, and local ponds and feeding them eggs. No infections resulted.[12]

The codex field notebook contains page after page of these three scientists' efforts to produce infections in any and every kind of snail or other aquatic invertebrate they could bring into the lab from Lake Texoma, the creeks that drained into it, and ponds that could easily overflow and dump residents into the big lake. All these efforts were failures. It eventually became obvious that none of the common snail species was serving as a required host for the worm's development. In the hands of Self and Peters, the worm could not get past its first embryonic stage. Fish with worms in their ovaries, releasing massive numbers of worm eggs into water where there were abundant snails of several different species, none of which became infected with this trematode, tells us that the life history of *N. texomensis* is nothing at all like we expect it to be.

That lesson of reality being the opposite of expectations gets applied easily to situations far beyond parasites in Lake Texoma fishes. My daily news feed is filled with examples of this lesson at work, especially with respect to some of the hot-button issues of our time—abortion, immigration, racism, and LGBTQ+ rights, all highly complex phenomena seemingly simplified by their single-word descriptions. Anyone who delves into these issues in detail, especially if driven by a desire to help resolve problems stirred up by political rhetoric, quickly discovers that the reality of individual cases is highly diverse, driven by many different forces, spread across socioeconomic layers, and not nearly as simple as some candidates for public office might want you to believe. Substitute the word "textbook" for "candidates for public office" in the previous sentence and you have a description of the life of *N. texomensis*. Although molecular biologists have identified some intermediate hosts for related worm species, we still don't have a clear picture of the *N. texomensis* life cycle, completed with experimental infections.[13] Thus, science has made about as much progress on this parasitological problem as the politicians have made on those much larger ones.

Eventually, Self's curiosity about the life of this worm evolved into my own curiosity about the life of Clarence Edgar Davis, or at least about the part of it that intersected so mysteriously and briefly with mine. He was a person I had nodded hello to in the halls of that science building and who was notable because of his job and the color of his skin. Even in the summer of 1960, when I returned to the University of Oklahoma (OU) to pursue my dream of being a college prof, I was experienced enough socially, politically, and biologically to understand the challenges he faced. But that experience was not educational enough for me to do the one thing I wish today

I'd been perceptive enough to do, namely talk to Davis about his life in Louisiana, his reasons for being at OU, and especially his reasons for being employed by Dr. Self.

Diaries are notorious sources of historical perspective, and the Self codex is clearly a special sort of diary. It's impossible to read this set of field notes without constantly being reminded that science is conducted within a social context, which means that the scientists doing their work are also embedded in this cultural environment. That situation has not changed since Davis came to work at the University of Oklahoma, and if anything, the cultural environment has become increasingly difficult to negotiate. Rarely a day goes by that news sources fail to report a conflict involving education, usually resulting from an infectious idea that something being presented to young people, ranging from preschoolers to college seniors, is dangerous in an unspecified way. I'm completely convinced that had I been smart enough to engage Davis in some extended conversations, his experiences and stories would have shaped my behavior as a college prof throughout my fifty-year career.

The people who spent so much effort on *N. texomensis* have shown us just how strongly curiosity about natural phenomena can drive a scientist's behavior. They've also demonstrated that persistence is a fundamental behavioral trait of scientists, one that can easily be expressed even in the face of ongoing failure. Would Dr. Self and his students have continued to pursue their research on this parasite if everyone they encountered told them they were fools? My guess is the answer would be "yes" and would be based partly on the fact that no matter what kind of scientific research one does, that work teaches us how science is done and what is involved in making the kinds of discoveries that eventually benefit us. The general rule is that science produces scientists, thus the benefits we all enjoy from scientific research are secondary products. That's a life lesson that elected officials would be advised to learn.

But the big take-home lessons from the *N. texomensis* story are as follows: *Seemingly simple problems can have solutions that defy the finest minds; expectations guide the search for truth, but they can also hinder that search;* and *nature doesn't necessarily behave the way you want it to or believe it should.* The last is, of course, true regardless of how much you want to deny it, how strongly some politician claims the opposite, or how hard you try to make it match your expectations. Those lessons also apply to a nation that has steadfastly denied what the scientific community has told it about climate change while being invaded by a virus enemy that killed more than a million of its citizens. And one part of nature that is not behaving as expected is a human mind that opposes use of the only weapon, for example vaccination, known to be effective against such a viral invader. I'm guessing that Clarence Davis could have predicted this relationship between nature and humans and probably delivered his prediction with the wry smile that I remember from passing him in the hall.

CHAPTER TWO

CHEYENNE BOTTOMS

The Global Reach of Parasites

> Spasskaya theorizes that the migrants are infected with different parasites in different geographical areas. My study was designed to test this hypothesis using a single species of migratory bird, the Pectoral Sandpiper, *Erolia melanotos*.
>
> Dan Rae Harlow, 1963[1]

About seven miles northeast of Great Bend, Kansas, lies one of North America's most impressive wetlands, sixty-four square miles of marsh—largest in the United States' interior—with water averaging about a foot deep, supplied by Blood Creek and Deception Creek in wet years.[2] Mosquito populations explode with spring rains that flood grasslands bordering the marshes; migrant waterfowl drop feces full of tapeworm eggs from worms picked up in Texas; pelicans sometimes eat so many carp they struggle to fly. Water snakes grow thick; turtles thrive; thunderstorms spawn lightning, massive hailstones, and tornadoes that sweep through small communities in nearby counties. On their way to northern breeding grounds, half a million migrating shorebirds gorge on invertebrate fauna supported by decaying vegetation. Kansas sunrises and sunsets over a flat horizon strain your sense of wonder. The whole place smells like biology. Bird migration makes this wetland an ideal place to discover how parasites move between continents, but you need to look inside a bird to find its worms.

What makes the Cheyenne Bottoms also a bird-watcher's paradise is a series of dikes, topped by gravel roads that allow you to drive out through the cattails, rest a spotting scope on an open window, and watch a dozen different species go through their daily routines. Of the 471 species of birds recorded for Kansas, more than 300 have been seen at "the Bottoms."[3] Out on the dikes, you can roll down the windows, listen to the wind and avian orchestra, smell the water, and welcome some of the world's more beautiful mosquitoes into your vehicle. Something about this place makes you want to admire the mosquitoes, at least if you're a parasitologist; they've survived the Kansas winter. You know they transmit malarial parasites to the birds and hope they don't also deliver some encephalitis viruses to you, viruses carried by sandpiper wings from Argentina instead of in an airplane from China.

Depending on the time of year, shorebird migration can be particularly dramatic. In the spring, long-distance migrants such as pectoral sandpipers bring flukes and tapeworms from South America and shed worm eggs into the marsh; in the fall, those same birds shed eggs from worms picked up in their northern nesting grounds. For someone studying the intercontinental movement of parasites, these dike roads provide an ideal platform for poking a shotgun out the window, collecting migrants, identifying their worms, and thus testing hypotheses about the spread of infectious agents between continents. People who do this kind of research are never surprised when other germs, including viruses and destructive ideas, invade a nation. They also use some of the same mathematical equations—analytical tools to handle data—used by those studying the movement of cultural items, for example smartphones, through a population.[4]

The first person I met who was doing research on the intercontinental movement of infectious organisms was Dan Rae Harlow, a grad student at the University of Oklahoma in the early 1960s, working on a master's degree under the supervision of Dr. J. Teague Self, the parasitologist obsessed with worms in a buffalo fish's ovaries. Because my wife, Karen, was Dr. Self's secretary, we were acquainted with Dan. I was also a grad student at the time; my curriculum included a class in comparative physiology, and he was in that class. I don't know why he started coming over to our small house, usually bringing a bottle of Chivas Regal, but when he showed up, that was Karen's excuse to make chocolate chip cookies. After the three of us finished discussing local politics and department intrigue, Dan and I would continue the conversation mostly about the physiology of crabs and marine worms. But eventually, as happens in graduate student circles, talk turned to our respective thesis projects—problems, successes, progress, and what we thought our chosen subjects were teaching us about life on Earth, including our own.

Usually all we could say about this last issue was that we were satisfying our own curiosity about how the world really worked; we were behaving like scientists being taught life lessons by our subjects. And our projects became our identities, at least while we were working on them. Dan's project at the time was a remarkable combination of conceptual power and audacity. He was trying to answer two simple questions: Do migrating birds bring parasites such as tapeworms, roundworms, and flukes into the Northern Hemisphere from South America?[5] And do migrating birds leave their nesting grounds in Canada with a characteristic group of parasites that they then transport to South America?

The significance of this project was obvious to us both, although at the time, it was relevant mainly to parasitic worms. "What is it?" is the first question every biologist must answer when encountering nature, and we all know that different regions of the planet can have quite different flora and fauna. If migratory birds are distributing their parasites across the hemispheres, that fact must be considered when trying to identify specimens or describe new species, thus adding to our inventory of life on Earth. But if the parasites stay in North America, for example, while their hosts migrate south and vice versa, that information also helps us understand the processes by which infectious agents are spread. Sixty years before that virus known as SARS-CoV-2 arrived on our shores, Dan Harlow was exploring the intercontinental movement of infectious agents, and his thesis questions were the driving force behind such exploration—simple questions with powerful answers applicable far beyond the issue of tapeworms in sandpipers.

By answering the first question, he could discover if those birds lost their North American parasites during the winter months in South America; by answering the second, he could discover if birds lost their South American worms while nesting in North America. Most of these worms require intermediate hosts, such as insects and crustaceans, to develop in before infecting birds, so the results would reveal whether those Southern Hemisphere invertebrates were satisfactory hosts and vice versa. Dan's work would thus reveal the mechanisms by which parasites survived migration. To address these questions about the intercontinental movement of infectious agents, he shot 124 pectoral sandpipers, now named *Calidris melanotos*, 81 in the fall and 43 in the spring; dissected them; collected worms from their intestines; and then went through the arduous process of identifying those parasites, a task that also involved laborious preparation of hundreds of specimens in a way that allowed him to see the diagnostic features. He knew that late nights in the lab would reveal what happens annually across the length of a hemisphere, events that had likely been happening for at least a million years.

Why did Dan choose pectoral sandpipers? Because they migrated the required distances between continents. In other words, nature did something interesting, so all you had to do was observe what was happening in nature and you'd end up with a really nice thesis that was also a major lesson in geography, ecology, ornithology, parasitology, history, the art of specimen preparation, the practice of finding and interpreting difficult scientific literature—some of it in obscure journals and a lot of it in foreign languages—English composition, and demands of publication. That diverse set of endeavors provided a rich repertoire of transferable skills, acquired by experience, in addition to the truly global perspective lesson taught by worms in sandpipers. Because of this research, Dan would never be surprised when a virus such as SARS-CoV-2 spread across the globe. Intercontinental movement of infectious agents, regardless of their type, would seem natural to him. Being a parasitologist, he'd also know that transmission mechanisms were a key to understanding such movements and to controlling their spread. I've not talked to him for at least sixty years, but I'd be very surprised if he didn't wear a mask and get vaccinated against COVID-19.

Compared to Dan's thesis project, mine seemed rather mundane, but it was eventually published. Thus, when it came time to pursue doctoral work, I decided to study parasites of birds. I wanted what he had acquired through his research—those life lessons on global affairs—so I approached Dr. Self; he agreed to take me on as a doctoral advisee. I had never taken a course in parasitology, however, so that summer, following receipt of my MS degree, Karen and I moved to the University of Oklahoma Biological Station (UOBS) on the north shore of Lake Texoma where Self was teaching parasitology. At UOBS, I spent eight weeks collecting birds, extracting their parasites, studying their blood smears, and learning the fine art of making tapeworm specimens. The smears were easy if I had fresh blood, putting a drop on a clean glass slide, putting the edge of another slide on the drop, and pulling the layer of blood the length of a slide. Tapeworm specimens were a different matter, requiring careful removal from an intestine; fixation in formalin, alcohol, and acetic acid; staining with a dye; dehydration with alcohol; replacement of alcohol with an organic solvent; replacement of the solvent with a resin; and careful arrangement of worm pieces on a slide. Compared to blood smears, tapeworm specimens are works of art.

I also identified my first mosquito using scientific literature in the station library—the notorious malaria vector *Anopheles quadrimaculatus*, gently captured while it was feeding on me—and found my first blood parasite, a species of the malarial genus *Plasmodium* living in the red blood cells of a meadowlark. I will admit that my heart rate went up with this first observation of an infected cell. With those two events, my mind was made up: for my doctoral dissertation, I would study malarial parasites of birds and get acquainted with mosquitoes. Malaria would be the vehicle I'd ride into my future as a scientist. I hoped that malaria in meadowlarks and starlings would give me the transferrable skills that Dan Harlow had acquired just by doing his research—the ability to use foreign-language literature, draw scientific diagrams, interpret complex literature, write coherent papers, talk to audiences without fear, and teach others about my discoveries.

So naturally, I immediately drew a picture of those infected blood cells, one with a multiplying parasite and others with the stages that would become gametes when sucked up by a mosquito. Cell (a) is an uninfected red blood cell; (b) has a malarial parasite that is multiplying; (c) has a parasite's "female" gametocyte that will become an ovum and be fertilized if it ends up in a mosquito's gut; and (d) has a parasite's "male" gametocyte that once inside a mosquito's gut will split up into sperm-like gametes that fertilize the ova.

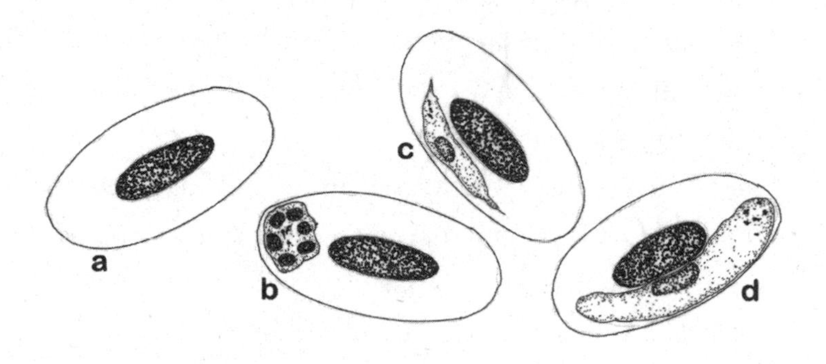

About this time, Dr. Self, in cooperation with a virologist at the University of Oklahoma Medical Center, Dr. Vernon Scott, and an ornithologist, Dr. David Parmelee from Emporia State College in Kansas, received a large grant from the National Institutes of Allergy and Infectious Diseases to do a multifaceted study of the intercontinental movement of parasites and viruses carried by migratory birds. The viruses were ones capable of infecting humans and livestock, so the work was considered important enough to merit funding. Dr. Self and I had discussed possible doctoral projects, particularly ones involving malarial parasites, and as a result, he offered me a position as research assistant on the Bird/Virus/Parasite project, which we started referring to simply as BVP. The main study site was the Cheyenne Bottoms, so for three years, I commuted three hundred miles between Norman, Oklahoma, and Great Bend, Kansas. I also had access to university vehicles, one of which was a 1957 Ford with the trunk replaced by a small pickup-type bed, in which I lost a couple of large water snakes, likely discovered by a later user. A second vehicle—a newish, 1964 Ford station wagon—was shattered by hail on one of the dike roads as another research assistant and I cowered in the back seat, hoping that a nearby tornado would not send us to the Land of Oz. As a result, I acquired the first of my research vehicle stories, luckily played out in Barton County, Kansas, instead of somewhere like central Mongolia.

The three scientists hired a full-time field man, Homer A. "Steve" Stephens, to collect birds, make blood smears, and put the carcasses in a deep freeze for later dissection. Bird species to be collected involved both long-distance and within-continent migrants. The final list was lesser yellowlegs (*Tringa flavipes*), both long- and short-billed dowitchers (*Limnodromus scolopaceus* and *L. griseus*), American coots (*Fulica americana*), red-winged blackbirds (*Agelaius phoeniceus*), starlings (*Sturnus vulgaris*), and western meadowlarks (*Sturnella neglecta*). Yellowlegs and dowitchers are medium- to long-distance migrants, wintering in southern states, Mexico, and South America; coots, blackbirds, and meadowlarks are medium-distance migrants; and starlings can be resident but also move south in the winter to varying extent.

The project's collector shot a whole lot of birds, all of which I eventually dissected. I came to know the insides of these bird species in great detail. The coots were full of remarkable tapeworms that for some mysterious and evolutionary reason had only two testes in each of their many segments. If you know tapeworms, you know that most of them have a whole lot of testes and a whole lot of segments, as well as ovaries, so they are reproductive machines. Life is chancy if you're a parasitic worm, hence the reproductive output; counting testes in tapeworms is a lesson in what it takes to be successful in an unforgiving world.

Nobody knows why the worms from coots have only two testes. That's exactly the kind of evolutionary question that could function to control a person's life, not unlike *Nematobothrium texomensis* had done years earlier for Teague Self. That's also a lesson in comparative zoology, taught by parasites. From going through the literature and worms from the other host species, trying to identify them and decide whether they might be new to science, you end up not only counting testes but also making sketches of the other organs—ovaries, glands that supply tapeworm egg yolk, glands that make the egg shell—being constantly reminded of the two guiding questions in biology: In what way(s) are these organisms similar in spite of their differences? And in what way(s) are these organisms different in spite of their similarities? Those two questions, easily acquired by those who study tapeworm and fluke anatomy, are tattooed on most biologists' brains for the rest of their lives regardless of their study subjects.

In addition to the research on worms and malarial parasites, another key component of this BVP study was the potential for movement of equine encephalitis viruses, which are transmitted by mosquitoes and are capable of infecting humans, and which birds were suspected to carry across continents.[6] Thus, part of the work involved mosquitoes, which I collected in what seemed like astronomical numbers at the time, mainly using light traps with a battery-operated bulb and a fan to blow any insects attracted to the light down into a cage. I also used dry ice traps because mosquitoes are attracted to carbon dioxide, which mammals breathe out, and that's what dry ice is: frozen CO_2. These latter traps were five-gallon containers with one end replaced with an inward-facing screen cone; mosquitoes followed the trail of CO_2, flew into the cone, and couldn't get out. Once collected, either the light trap cage or the carbon dioxide trap was put into a freezer, killing the mosquitoes. I then delivered them, frozen, to the University of Oklahoma Medical Center, where I sorted them into species and Dr. Scott tested them for viruses. I eventually became reasonably well educated about mosquitoes but, more importantly, enthralled with their beauty under a microscope.

That mosquito beauty, recognized because of how I was looking at them, why I was looking at them, and the kind of relationship they had to my career, was almost like a big slap in the face, a major wake-up call about the lessons taught by our research subjects. I would get up from that microscope, walk to my car, drive through the Oklahoma City neighborhood surrounding the medical center, and suddenly the city looked different than it had during my early years of shooting sparrows with my BB gun and trapping muskrats on the country club golf course. I looked for places where mosquitoes might be breeding, wondered what species might be in this suburban environment, and thought about stopping people on the street and asking if they'd ever had a chance to look at one under a microscope.

Our fear of mosquitoes is derived from their role as transmitters of disease and the fact that they can deliver an itchy welt. My newly found love of mosquitoes was derived from knowing them for what they were, from seeing firsthand the differences in various species' breeding biology and having to study their different wing veins and color patterns to identify them. Because of this work, I became very well acquainted with organisms that most people kill without ever knowing or caring about them. Mosquitoes thus taught me that how and why you look at a stranger, along with whatever intellectual or cultural baggage you bring to the encounter, shapes your reaction to that stranger. Another life lesson from parasites, this time from ones that will suck your blood if you let them.

Of all those life lessons acquired from my research subjects, this one, relative to strangers and perceived danger, may be the most important. I know it has shaped my interactions with situations and people, including the thousands of students in my large classes over the years. Today my news sources are filled with examples of fear and rejection, mainly of the unknown and unfamiliar, exhibited by people in various positions of power, when caution and rationality are probably more appropriate responses. In these cases, the unknown and unfamiliar can range from scientific conclusions about climate change to the sexual orientation of neighbors, the impact of poverty on American citizens, and reproductive choices of women, just to name a few. Whatever intellectual and cultural baggage I brought to my encounters with Cheyenne Bottoms mosquitoes, that baggage was quickly abandoned when I entered their environment and studied them carefully and objectively, as legitimate inhabitants of my world. Yes, I saw nature's incredible beauty instead of, or maybe in addition to, bloodsucking flies just trying to find their next meal.

That lesson learned from collecting mosquitoes in a certain way and for a certain reason is still with me. Now, sitting on our patio in the evening, I try to catch the ones that land on my arm. I scoop up mosquito larvae from our birdbath and save them in a used prescription vial with the intent of trying to identify them. And that behavior extends to other potential disease vectors. Once, on a tourist safari in Tanzania, a tsetse fly, transmitter of deadly African sleeping sickness, landed on my hand. I slowly reached for my camera, careful not to disturb the fly, and was able to capture a perfect image of that famous vector feeding, my blood coursing up through its proboscis. Later, at an auction to benefit students studying parasitology, a bidder paid $450 for a copy of this photograph so he could use it as classroom material. With that money, some young scientist will get to attend a scientific conference, meet their heroes face-to-face like I had done as a student, and spend a week listening to the life lessons parasites have taught other scientists—all because a tsetse fly landed on my left hand, my right hand held a camera, and I'd learned to love mosquitoes fifty years earlier.

Did the virologists ever discover whether Cheyenne Bottoms mosquitoes were transmitting viruses from migratory birds to local ones? The specific act of transmission was never demonstrated unequivocally, but the potential for that happening was clear from work done at the University of Oklahoma Medical Center. Suckling mice, just a few days old, were injected with homogenized mosquitoes and blood samples from various migratory birds. If the mice died, the injection was considered to contain viruses; if the mice lived when the injected sample was treated beforehand with antibodies specific to various encephalitis viruses, the infective agent was tentatively identified.[7]

The virology results were published in 1965, in a paper titled "A Preliminary Report on Isolations of Arboviruses from Mosquitoes and Migratory Birds." The authors were Dr. Scott and two of his students, Mary Jo Hellmer and Roy McGuire. Yes, the mosquitoes contained twenty-seven different agents that killed the mice, and they found another nineteen in blood samples from migratory birds. Western equine encephalitis, infective to humans, was among the identified viruses.[8] The three scientists processed more than a hundred batches of my mosquitoes, including thousands of individuals, but they'd have had to come to Kansas and walk the marshes to truly appreciate the beauty of those vectors transmitting their viruses. The life lesson is obvious to a biologist: you're just one more batch of culture medium to a virus transmitted by mosquitoes, so if you get infected, the symptoms are a reminder that you're just a part of nature. Alternatively, of course, mosquitoes can end up teaching us that it's possible to love, or at least appreciate and understand, something others hate and fear.

My dissertation part of the BVP work involved studying blood smears, hundreds of them, from those various bird species, rearing mosquitoes and feeding them on birds to see which species could transmit the parasites, and analyzing my data. The birds I used in the lab were starlings, nestlings taken from tree holes high up in dead cottonwoods around the Cheyenne Bottoms marshes and reared in our kitchen. Needless to say, I fell in love with starlings too. Those hand-reared nestlings were the only birds used but not killed as part of the BVP project. They grew up and fledged in the same apartment as our daughter Cindy, age two, was also learning to fly in her own way. When they were no longer useful as experimental animals, I let them go, watching them fly out between a couple of OU buildings. I had faith, probably misplaced, in their ability to adjust to the wilderness of Norman, Oklahoma. Cindy also grew up and eventually flew away, adjusting to a much larger and varied wilderness than the town in which she was born. Now, decades later, starlings on the suet block outside our kitchen window bring back those times, at least in my mind. I marvel at their beauty and take their pictures, but they are generally considered invasive pests and not protected by federal law.[9]

Lessons learned from wild animals are never forgotten, especially when, as with mosquitoes and starlings, those wild animals are not always welcome in your neighborhood. Those lessons are easily applied to other creatures, including certain members of the diverse human population that now occupy the town where I live, sending their children to classes that I taught, and who are referred to regularly in our city council meetings and political campaigns but without the honor of being mentioned by Shakespeare as were starlings. Every time I read about local efforts to defeat proposed nondiscrimination ordinances, I'm reminded of this love for mosquitoes and starlings even though the political targets can be members of the LGBTQ+ community, immigrants, or ethnic minorities. Many years have passed since I last collected mosquitoes and starlings at the Cheyenne Bottoms, learning to respect them for what they are and what they could teach me, so that's long enough to call it a life lesson.

I spent many hours with mosquitoes and starlings, handling them, feeding them, and in the process learning about them in ways that many of my neighbors would never understand. The starlings were especially instructive—adaptable, amazingly vocal at times, and beautiful in their post-breeding plumage. That description is in sharp contrast to the official one—invasive and unprotected—rather similar in this regard to multitudes seeking refuge in this nation. Those human immigrants who have managed to find a nest in my community, however, have enriched that community in many ways, quite analogous to the way starlings did for me personally so many years ago. The ethnic food trucks, restaurants, shops, and houses of worship are just some of these contributions; their children's academic successes, as sometimes featured in my local newspaper, are impressive; and nowadays, it's normal to encounter them in any role from service personnel to physicians. If someone had told me back in the 1960s that birds and mosquitoes would shape my reactions to local and national politics sixty years later, I would never have believed them, but they would have been correct.

The work I did as a research assistant involved dissecting all the carcasses, harvesting the worms, preparing them for identification, and bleeding sentinel flocks, the latter consisting of three cages with ten chickens each, which were placed in strategic places around the marshes. In theory, if encephalitis viruses were being carried into Kansas by migratory birds, those viruses would be picked up by local mosquitoes and transmitted to local birds. Sentinel flock chickens were surrogates for the resident wild bird species; their cages and relative domestication meant, of course, that they were far easier to use in this transmission assay than wild birds would or could have been. So once a month from March through October, I would get in a University of Oklahoma vehicle, drive like the wind to the Cheyenne Bottoms, bleed the flocks, collect thousands of mosquitoes, retrieve carcasses of birds that Stephens had collected, and come home with blood smears, all in addition to whatever dissertation-related fieldwork I needed to do. My doctoral dissertation reveals that data set was extracted from 673 birds and 14,276 mosquitoes of various species.[10]

After all that collecting and dissecting, did Drs. Self and Scott answer the question about intercontinental transport of viruses and parasites? I don't know the final conclusions, but Kansas mosquitoes had encephalitis viruses as shown in that one publication, and the spring migrants had worms they had to have gotten on their wintering ground. I, however, got one hell of an education delivered by worms, mosquitoes, birds, and a prairie marsh. All the data and records, as well as thousands of parasite specimens, now reside in the University of Nebraska State Museum research collections, housed in the Harold W. Manter Laboratory of Parasitology. Under the right circumstances, a bright young person could walk into that lab, dig into those records and specimens, and emerge five years later with a rather remarkable doctoral dissertation on the intercontinental movement of tapeworms and other parasites based on field and lab work and all that shooting and dissecting done at the University of Oklahoma during the mid-1960s. A natural history museum thus houses a massive data set just waiting for the right personality, the right exploring mind, to find and use it. That's what natural history museums do, all around the world.

And what are those data likely to reveal? The answer requires an appreciation of parasite life cycles and the way evolution has shaped developmental events. In my admittedly biased view, the most beautiful worms in all those migratory birds are the tapeworms and flukes. When prepared properly and mounted on microscope slides, their internal anatomy is seen as complex, exceedingly diverse, but with recognizable elements—ovaries, testes, ducts, yolk glands—in species-specific arrangements. Both groups also need to pass some time in hosts other than sandpipers to develop properly and survive. In the case of tapeworms, those places are likely to be small crustaceans or aquatic insects; with flukes, those other places are snails and also, potentially, aquatic insects or crustaceans.

These invertebrates are sandpiper food, many of them intermediate hosts carrying worms, thus transmitting infections to a host that will fly across national boundaries. Every year that it's alive, every single bird, especially species such as pectoral and Baird's sandpipers that migrate such long distances, do what any invertebrate zoologist can only dream of—pick up samples across marine coasts and freshwater marshes, all along the sixty-five hundred miles from Argentina to Canada.[11] If the gut contents of just one of those sandpipers, invertebrates eaten on both northward and southern migrations, could be recovered and preserved, those partially digested prey items would likely support a doctoral dissertation and fill a small cabinet in some museum research collection facility. And if such a collection could be made from just one Baird's sandpiper and one pectoral sandpiper, from January to December of one year, it would likely take some beleaguered doctoral student ten years to sort through all of it, identify most of it, analyze its DNA, and describe all the new species discovered, having been eaten by two individual birds.[12]

If you collected a sandpiper from the Cheyenne Bottoms today and it was filled with adult worms, some of those would also go into ethyl alcohol for DNA sequencing. A month later, you might have the genetic information on those adult worms too, and with luck, you'd be able to match those DNA sequences with ones from larvae, thus connecting stages of a life cycle. In this way, we can sometimes discover parasite life cycles using molecular technology, not unlike solving a crime or paternity claim, thus answering the question of what that bird ate to get infected with that worm. But the travels of an individual bird are generally unknown and unknowable, except for distance as revealed by band numbers. We can never learn exactly what happens to one wild sandpiper between the time it gets banded and the time its band number is read. For this reason, that mythical collection of invertebrates consumed by just two sandpipers can never be assembled; it's just a figment of our imagination, based only on what we know about sandpipers in general. Our studies of parasites in Kansas birds teach us that there are some things we can never discover, a life lesson that lasts far beyond our graduate student days and applies to situations far beyond tapeworms inside birds shot in Kansas.

What we do know, however, is that as long as they are alive, worm parasites produce eggs that get shed in a host's feces. The parasites have no control over their lives, except to produce massive numbers of eggs and live for a certain length of time. Their evolutionary history locks them into this life dependent on various species for sustenance and the environment necessary for embryonic development. That history can also put them in sandpipers migrating between continents instead of in January chickadees at your snowbound backyard feeder. The requirements for worm development, as strict physiologically as those of a human embryo, are fixed by this evolutionary history. In practice, their lives are subject to constant change, although this change is predictable—Earth circles the sun and its axis is at an angle, thus producing the seasons. Sandpipers migrate annually and have for millennia, along the way eating the requisite species of insects, crustaceans, and snails that lived in preindustrial environments stable for thousands of years.

The worms handle predictable change by adapting in the only way they can: numbers of eggs and longevity. If change is predictable, species can adapt, but only by using their genetically endowed traits and inherent variations expressed over generations, allowing some variants to reproduce more successfully than others. If there ever was a lesson taught by tapeworms and applicable to humans living on a rapidly warming planet, this is it: adaptation, not denial, is the key to survival. This planetary process affecting sandpipers, tapeworms, and humans has manifested itself numerous times over the millennia, but not until the twentieth century have some members of a species been both willing and able to deny it, a case of purposefully ignoring what the planet is telling us. Eventually, of course, we will end up exactly like those sandpipers and their worms, either adapting to the forces we cannot control or suffering the consequences on a global scale.

Now, after half a century, all that fieldwork, collecting, dissecting, specimen preparation, and publication is assembled into my professional and personal existence. My career was built from those birds, worms, and mosquitoes, and as a result, so was my ability to provide for a family. Today, as a senior citizen, when I reflect on life, family, and career, that combination of Dan Harlow's thesis, my wife's job as secretary to Dr. Self, and the BVP project seems like a fortuitous set of events in which I was simply caught up, dragged along, and rewarded, ultimately and quite generously by the University of Nebraska, for collecting animals and studying their parasites.

It's those lessons about fixed systems in changing environments, however, just like parasite life cycles being carried across the Gulf of Mexico into central Kansas, that spill over into my nonparasitological thoughts. In those thoughts, the fixed systems are human ones: behaviors, desires, conflicts, beliefs, loves, hates, and fears, all being played out in a cultural milieu that technology is altering—perhaps beyond our control—at an unprecedented rate, even as our physical environment is also being dramatically altered by planetary forces that are most certainly beyond our control and even as we humans are perpetrating another mass extinction by our fecundity and our environmental destruction. It's not surprising that I cannot look back on that research so tied to migration without imagining, or at least trying to imagine, the journeys that human migrants are making today, largely a result of forces beyond their control.

The most impressive biology lesson of our times, however, also involves the intercontinental movement of an infectious agent, in this case a virus, transported by humans as the vectors, our movements aided by flying machines. SARS-CoV-2 slammed into fixed systems all around the world, with some responding appropriately to the best of their ability—adapting—but others denying that the change had occurred or at least acting as though the change was inconsequential, and still others trying to educate the public about threats and solutions, both real and imagined. This diversity of responses to a virus mimics that of established cultures to perceived human invaders, even when the differences between residents and newcomers are matters of language, facial features, skin color, and cultural practices instead of deadly disease. Research on this issue reveals that the main concerns driving reactions are perceived economic and cultural threats, the latter often involving social practices and religion.[13]

Thus, we're learning how human behavior driven by parasitic ideas, beliefs, and at times outright ignorance can promote the spread of not only a virus but also more ideas, beliefs, and ignorance, compounding the impact that infective language can have on a host population and driving change that sometimes is quite unwelcome. At the Cheyenne Bottoms, it seemed natural for birds to move between continents, carrying their parasites, dropping worm eggs in their feces, and encountering various environments along the way. Nowadays, it seems just as natural for people to move great distances, dropping money, goods, services, and cultural items along the way. Globalization is a modern fact of life as surely as tapeworms and flukes have been a fact of sandpiper life for millennia. I can just imagine the conversation we would have this evening should Dan Harlow show up at our front door with a bottle of Chivas Regal. There would be no kind words for political "leaders" who have not learned this life lesson.

CHAPTER THREE

MEADOWLARK

Beautiful Host, Armed Parasite

The lark that shuns on lofty boughs to build
Her humble nest, lies silent in the field.

Edmund Waller, *Of and to the Queene*, 1645[1]

In the morning, I remove a bird from the freezer. Later that afternoon, after it's thawed, I take it out of its plastic bag and read the tag attached to its leg: 2549–50. Its feathers are damp from having been frozen in dry ice months earlier, transported three hundred miles from central Kansas to Oklahoma in an ice chest, and transferred to a deep freeze. As always with these birds collected for worms, I stroke the feathers, especially those on the wings, smoothing them in places, preening them with my fingers to close gaps in the barbs. I was shown how to do this preening by an ornithologist who shot birds, skinned them, and preserved the specimens, almost like works of art. His specimens were deposited in a museum to demonstrate beyond doubt that a species occurred in a particular place on a particular day—an indisputable fact in a world awash with fiction spread across land and sea at the speed of electrons.

If there are worms in number 2549–50, they too will find their way into a museum, tangible evidence of an infection. In a so-called information age flooded with conspiracy theories, fake news, and internet blathering from a billion sources, one bird's skin and the worms held within are irrefutable proof of what the world was really like in one place, at one time—a literal truth preserved for as long as the museum remains intact. And if the blood smears from this bird show an infection by malarial parasites, a Kansas meadowlark will be the teacher, I will be the student, and the lesson will be about mosquitoes and malaria, a disease that still ravages people throughout the tropics. Even though the species of malarial parasites in birds are not infective for humans, the lessons they can teach us about transmission, diversity, diagnosis, identification, and the critical importance of tangible evidence such as a blood smear in the museum are of lasting value when we encounter another infectious disease, regardless of the agent causing it or the place on Earth where we encounter it.

Sturnella neglecta is the western meadowlark; dark bars across the wing and tail feathers are not confluent along the shafts, a color pattern that tends to make these birds appear lighter than their eastern cousins, *S. magna*, a difference interpreted as an adaptation to the arid plains.[2] Once my admiration for the beauty of this specimen is satisfied, I dig into its body cavity with my fingers, extract the entire digestive system, and drop the guts into a pan lined with paraffin. When this one was shot months earlier, a hired collector immediately cut open its chest cavity, extracted the heart, and smeared some blood on two glass slides before slipping a numbered and dated tag on its leg, then dropping the whole bird into a plastic bag. The numbers on the tag matched those on the slides. The bag went immediately into a container with dry ice. Months later, guts in hand, I start looking for worms that lived and died in this meadowlark, asking myself what I can learn about infective agents from whatever I find along this journey through an intestine, knowing that the blood smears don't require immediate attention like the worms do but also knowing that those smears will introduce me to Kansas mosquitoes as disease vectors instead of as just annoying pests.

I've lost count of the meadowlarks that I've handled this way, but that total is easily retrievable. As with all the others, that numbered tag on its leg distinguishes an individual from its kind. Even as I put the entire mass of intestines in a wax-bottom pan with water, pulling the gut apart from the mesenteries that keep it folded and looped inside a living bird, I picture the scene in which this one was killed for its parasites. It's the last day of July 1964; the temperature is 102°F. A collector, paid to shoot birds, drives down a gravel road near a vast marsh. His truck is not air-conditioned. He's a chain-smoker; an ever-present cigarette, a Salem, held by lips weathered by the plains wind and sun, drops ashes onto his leg. He doesn't care about ashes. He doesn't care about the mosquitoes swarming in from the nearby ditch. All he cares about is one meadowlark in the middle of Kansas.

A hundred yards ahead, this *S. neglecta*, now known as 2549–50, rests on a bois d'arc fence post. The man slows; the bird still does not fly. The collector reaches slowly across the seat, picks up his .410 shotgun, aims out the window, and pulls the trigger. A few seconds later, 2549–50 has its sternum ripped open, its heart cut out, and its blood smeared as thinly as the man can manage on the two glass slides; another thirty seconds and it's tagged, bagged, and dropped into a chest loaded with dry ice. Only then does the man pay attention to the mosquitoes on his arm and neck, the ones that may have been biting 2549–50 before or even while it was being killed. After the mosquitoes, he deals with the cigarette.

The number on this meadowlark's leg tag is the same as the number scratched with a diamond-tipped pen on those glass slides. The collector knows that sometime soon, probably within a month, I will get in a car and drive the 288 miles from Norman, Oklahoma, to a small wooden building at the Cheyenne Bottoms, that 64-square-mile marsh northeast of Great Bend, Kansas, and pick up this meadowlark, along with the others he's collected and a box of blood smears, then deliver the cargo back to Norman, Oklahoma, where I'll finish looking through intestinal contents before turning to the smears, hoping to find parasites within cells. The marsh is a wonderland of mosquitoes, some of them stunningly beautiful under a microscope. Surely one of them bit this bird and transmitted blood parasites; that is my expectation if not my hope. In my mind, the bird in my hand is a model for one of the world's truly important diseases—malaria.

Meadowlarks and mosquitoes will teach me about malaria in ways that I could never learn from a textbook. Meadowlarks and worms will teach me about the value of tangible evidence in the coming information age awash in misinformation. But it was only decades later that I realized what I'd learned by studying all the parasites in meadowlarks and comparing those lessons with what I'd learned from starlings collected and handled in the same way as 2549–50. Comparison of infections in different species raises questions about the reasons behind any differences. Do starlings and meadowlarks have the same species of malarial parasites? Do they get bitten by the same species of mosquitoes? Do their infections develop in similar or different ways? By addressing these kinds of questions, birds end up teaching me how to study human diseases.

Months after that July 1964 day in Kansas, I hold 2549–50's visceral mass draped across my wet left hand, clip off the gizzard, letting it drop into the water, and slip one blade of small stainless-steel scissors into the opening. Slowly, making sure the scissors blade is pressed as closely as possible against the intestinal lining, I slice open the entire gut, then drop it into the water, teasing it into a spread-out ribbon. The intestinal lining is buffish gray; there are grasshopper parts left over from a meal snapped up half an hour before the bird's death. Anything white is a parasite. When you live inside another animal, there is no evolutionary advantage to flashy color; mates are not attracted so much as encountered by chance, depending on what that host animal has eaten.

What do I hope to find inside these guts? Worms. Thorny-headed worms. Tapeworms. Roundworms. Even as I gently scrape loose the intestinal lining, I wonder what it is about worms inside other animals that is so captivating, so able to change the trajectory of someone's career. A worm inside a meadowlark conjures up an image of a worm inside you, an image of harm, a kind of dirtiness not easily washed away, a hidden intruder no longer hidden because it's imagined, and the fact that it's only imagined is almost as effective as if it was really there. What would other people think of me, you wonder, if they knew I had this worm in my guts? Would they be envious, disgusted, curious about the character of a person who maybe didn't really care whether he had a tapeworm or not? Would they immediately ask how to get rid of this creature making your gut its home? What would they think of me if they knew I had pathological ideas in my head, much more dangerous than a tapeworm in my belly? Would they ignore the ideas and focus on the worm, or would they admire me because of those ideas? If they admired me because of dangerous ideas, would that make them more of a problem for my society than the tapeworm in my guts?

Yes, I am completely convinced that the allure of parasitology is closely related to our ability to envision ourselves infected. No one can prove what happens to a person's mind as a result of imagined parasitic worms, although it's easy to predict what happens when that same mind is infected with certain ideas. Infection changes behavior. We get sick and take drugs, an act stimulated by awareness of that foreign body suddenly living in ours. Annual flu shots are choices based on knowledge acquired from a variety of sources. Those same sources can deliver other infectious ideas, however, some of which spread through human populations, producing events we read about not only in history books but also in our daily newspaper. In the year I'm writing this paragraph, for example, the word "woke" produces reactions across the political spectrum and throughout social media, some using it as an expression of defiance, others as an expression of disdain, with subsequent behaviors manifested at the ballot box. That single word carries the idea of social and racial justice throughout a nation just as effectively as a thorny-headed worm egg carries an embryo into the pasture. Infection is a phenomenon far more general than most of us realize or like to admit. A worm in a meadowlark reminds me of this principle.

But we can also get infected with the idea of a career in science, envisioning a lifetime spent answering to our own curiosity about how phenomena move through populations, not only of meadowlarks but also of human beings. I've seen this set of events happen so many times that I honestly believe that my interpretation of them is close to correct. Twenty years after digging through 2549–50's guts, I'm a college prof, teaching parasitology and watching as a young person sets a trap: a small, elongated, aluminum box with a treadle in the floor and a door activated by this treadle. The trapdoor folds up, snapping down when an unsuspecting mouse follows the odor trail of peanut butter mixed with oatmeal inside, sniffing its way to death, and steps on the trigger. A meadowlark back then, a field mouse now; the fact that two people view them as carriers for worms ties them together across time and taxonomy. The fact that two people, a generation apart, view them and their parasites as teachers also ties them together across time and taxonomy. Infective things can be infective in more than one way, leading to many different outcomes ranging from fear to desire, from ignoring them to using them to build a career.

In the morning, our nascent parasitologist walks the wet meadow, finds the survey flag, and stops, her heartbeat picking up considerably at the sight of a sprung trapdoor. She collects the flag and the aluminum box, feeling the added weight and hearing something, scratching, maybe. Suddenly there is the realization that she must dissect this mouse and dig through it for worms. She gets a whiff of urine. Mouse fecal pellets will be inside, maybe containing worm eggs or other kinds of parasites packaged in capsules called "cysts," or deadly hantavirus particles. How do I do this, she wonders, without getting bitten? There is no paid collector with a shotgun to prepare this mouse for dissection, as was the case with *S. neglecta* number 2549–50. As she walks back to the laboratory, she realizes that she's about to become a parasite hunter looking for worms, learning firsthand not only about but also from the most common way of life among animals on Earth.

That scenario with a student and her first kill for worms is a product of my imagination after disemboweling 2549–50. Even as I slowly and carefully scrape off 2549–50's intestinal lining, gently teasing out the proboscis of one thorny-headed worm, then another, and making sure that a tapeworm's anterior end is intact, I'm picturing myself as a scientist, a parasitologist, for the rest of my professional life. Sometime in the future, I will ask a student to collect specimens to find some worms, just like my academic adviser, John Teague Self, did with me. *I want to be one of these kinds of scientists*, I think while using a camel's hair brush to lift both the thorny-headed worms into a petri dish of water, then doing the same with the tapeworm, before putting both dishes in the fridge, sitting back on my lab stool, and contemplating what I'd like to accomplish in the decades to come.

Hopefully those years will be spent as an academic scientist, a parasitologist. I want to spend the rest of my life sending someone out into the field to collect something and find parasites. I want that person to be captivated by the elegance of a properly prepared tapeworm specimen, appreciating how its suckers, maturing reproductive organs, and ducts have become a work of art with proper use of stains. I want that person to wonder how thorny-headed worms evolve their hooks, consider what happens to a host's gut when those thorns get embedded in it, and ponder the symbolism of a vaginal plug that prevents subsequent males from mating with a particular female. I want someone else to get infected with a desire to study the most common way of life among animals on Earth, even if that study involves killing something, including something beautiful like a field mouse, a meadowlark, or a mosquito.

This kind of dreaming and wishing consumes the hours before I take the dishes out of the fridge and, using the same brush, transfer these worms to a fixative—formalin, ethyl alcohol, and acetic acid, mixed in standard proportions memorized by every parasitologist who's ever dug through some guts while looking for worms. Over the next few days, those worms will be washed, stained, and mounted on microscope slides. Sometime in the future, 2549–50's tapeworm and two thorny-headed worms will end up in a museum, there to rest, cataloged, available for borrowing by scientists anywhere in the world, for as long as that museum and the civilized society that supports it exist. The Canada balsam in which these worms now rest is intended to keep them safe, embedded like Mesozoic insects in amber, for countless millennia.

From reading the paleontological literature and studying museum specimens made decades ago by other scientists, my imaginary story now leaps several thousand years into Earth's future. Some explorers from another planet are digging through layers of soil. One of them finds a glass slide with a thorny-headed worm mounted in Canada balsam. All the cities are gone, buried in carnage produced by wars, climate change, and a particular primate's inability to make its peace with the only planet known at the time to support life. Yet here, on this slide, is irrefutable evidence for what a prior world was really like. Something similar to a hand holds the slide, turning it over and over again.

"What is this thing?" asks one of our aliens.

"It looks like a parasitic worm," answers her companion. In another galaxy, before she entered that school for explorers, she'd been to another kind of school, hoping to become a physician, where her instructors made her cut up something that looked like a fish. She'd found a worm inside this water creature; the worm moved in strange ways. Her first question was how that worm got inside the thing she'd just dissected. Her instructor could not answer her question, so she became an astronaut who never forgot that she'd seen some long, thin thing living inside another thing that came out of an ocean on a distant planet.

My own hours in the lab watching thorny-headed worms lose their water, lose their alcohol, get saturated with an organic solvent like xylene, and get embedded in balsam are creative thinking hours. Like all parasitologists preparing specimens, I end up multitasking; one of those tasks is to imagine a world so far into the future that it's science fiction, a natural consequence of following standard preparation methods. But just like the balsam that trapped insects during the Mesozoic, that balsam into which I put this thorny-headed worm will survive not only into this imaginary world with that visitor from a galaxy very far away but also into the real world unless we destroy our museums, our repository for tangible evidence of what the world is or was really like.

This sci-fi scenario is no more far-fetched than some anthropologist finding a nearly complete prehuman's fossilized remains in an Ethiopian desert that used to be a lush tropical forest. Scientists who work in museums, who deposit specimens in museums, and who borrow specimens from other museums understand our need for evidence of what the world is really like. In contrast to the political rhetoric splashed across my daily newspaper or invading my social media feed, all of which is designed to make me *believe* something and act accordingly, museum specimens are silent reminders that the unassailable truth does indeed exist. Hooks on a thorny-headed worm do not try to convince you they are better than the ones on a tapeworm in the next box of slides. Nor do those hooks try to make you believe they belong to a different worm species. When scientists handle museum specimens, they understand that sometime in the past, another individual made every possible effort to preserve the literal truth about some small part of the world. That's probably the reason scientists can be so skeptical about claims made in the heat of political campaigns and scan social media for entertainment instead of advice.

Had my morning with 2549–50's intestinal worms happened in the present, with its stunning repertoire of molecular tools with which to explore all things biological, I would have saved a piece of the bird's breast muscle, one of its thorny-headed worms, and a piece of its tapeworm, all in 95 percent ethyl alcohol for later DNA analysis. Because meadowlarks, like birds in general, are really flying ecological communities, I would have looked through its feathers for lice and mites, saving some in alcohol and putting others on slides, processing them as I did the worms. A sample of gut contents would have been saved in a vial of sodium dichromate for diagnosis of a possible infection with coccidians, the same group of parasites that can wreak havoc on poultry operations.

Number 2549–50 would then have been skinned; its skin would have been rubbed with borax and cornmeal to discourage insect pests and remove fat, respectively, filled with sterile cotton, and sewn up, tagged, and dried, later to be deposited and cataloged, its numbers then entered into a database searchable by anyone in the world. But it was the fall of 1964 when my hands were covered with 2549–50's blood and gut contents. Back then, the molecular realm was a distant dream, as science fiction as my story of aliens finding that thorny-headed worm embedded in balsam. So 2549–50's carcass ends up in the landfill. I was too busy at the time, too preoccupied with my thoughts about malaria, mosquitoes, and a scientific career, to do that kind of complete examination and preservation. Nor was such a complete data retrieval and preservation demanded by the research protocols that are now required and enforced. Nevertheless, those thorny-headed worms did end up in a museum research collection, and today I'm going to see if I can find them.

The Harold W. Manter Laboratory of Parasitology, known on

specimen tags as HWML, sits atop a former Elgin watch factory building, now named Nebraska Hall, abbreviated NH on University of Nebraska–Lincoln (UNL) maps.[3] The lab is an administrative unit of the University of Nebraska State Museum; NH is three blocks away from Morrill Hall, the exhibits building that houses *Archidiskodon imperator maibeni*, known locally as "Archie," the largest of all woolly mammoth fossils, replicated as a bronze sculpture at the entrance. On a late fall day, I walk past the bronze Archie, smiling at schoolchildren trying to touch his uplifted foot, continue the three blocks to NH, and take an elevator to the fifth floor where my ID card trips the electronic lock and allows me into a long hallway. Framed photographs of parasitologists cover a wall on one side of the HWML entrance. I stop and review the images—friends, former students, colleagues, famous scientists I know only from scientific literature, some still alive, quite a few deceased, including J. Teague Self, the person who'd sent me to Kansas to get that meadowlark, its worms, and the blood smears made by a collector. Self also deposited my thorny-headed worm in the HWML collections, the very worm I'd recovered from 2549–50 in the early fall of 1964. With the help of Gabor Racz, the collections manager, I find it and do the typical parasitologist's behavior: draw its picture.

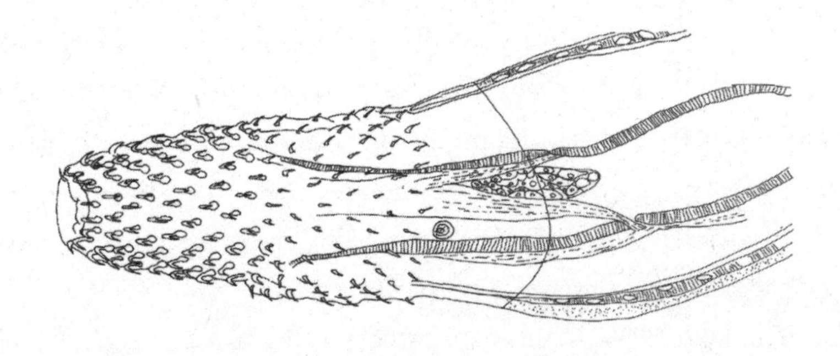

From those days sixty years ago, I remember a name—*Mediorhynchus grandis*—living in my memory not only from the work of other parasite hunters but also by the mystery of transmission, a mystery which, unlike that of the worm *Nematobothrium texomensis*, was solved. Thorny-headed worms in a bird's gut mean that something eaten by that bird had immature worms inside it, keeping development on hold until a beak crushed whatever insect held the parasite larva and a unique combination of digestive enzymes tripped the embryonic triggers necessary for excystment, growth, and ultimately, if the gods of chance are willing, sex with another worm. In 1962, a man named Donald V. Moore told us that the insect was likely a grasshopper that had, a couple of months before 2549–50 was shot, eaten an egg from *M. grandis*, shed in meadowlark manure somewhere in the Kansas prairies. Moore also told us that it took another month before that worm grew up and found another worm that had somehow by chance alone also ended up in 2549–50's intestine, at exactly the right time to encounter a mate.[4]

The chance of these eating events occurring in central Kansas, of the National Institutes of Health granting money for a collector to be hired and a graduate student to study the results, and of 2549–50 being killed for worms is so small that it cannot be expressed in an understandable way. But that is only the chance of one chosen bird being infected. The chance of Kansas meadowlarks in general being infected, however, is easily calculated if you dig through enough of them and record which ones have worms, which is exactly the kind of activity I'm involved in as I dismantle 2549–50. And because of this work of digging for worms, I come to a sense of what it means to be an individual versus a member of a species, any species, including my own. This ability, this willingness, to distinguish between an individual and its kind may be the most important product of the work I do. It's a habit formed then embedded in my brain like some parasite, influencing the way I interpret political rhetoric for the rest of my life, leading eventually to questions I cannot answer but cannot erase from my thoughts: What is a human being? And a larger one: What is *Homo sapiens*? And perhaps the largest one of all: What are we doing to define our species? The answer to that last question changes daily in some ways, even as, in other ways, it remains constant over historical time.

Yet here I am, with the help of a museum scientist, Gabor Racz, holding a glass slide with a worm from that bird on it, about to watch as he takes a photo of the thorny head. Racz's primary responsibilities are to ensure that museum records are accurate and complete, specimens are housed properly, the collections are secure, and users are vetted. Everything he does is designed to validate museum collections as accurate records of our world. His work is in stark contrast to some of the versions of "reality" displayed on social media and in the daily news. Racz is the equivalent of a highly skilled and objective journalist, but his subjects generally can't argue with what he reports. His use of the microscope and camera, however, are about to uncover some of my past behavior in the lab.

Sixty years ago, I did not handle this particular worm in an ideal manner, because its "head," which is actually an eversible proboscis, is partly retracted into its body, rather like a sock turned inside out, and now with some of the hooks inside. I should have left it in the fridge longer so that it would die and relax instead of having enough life left to pull that proboscis back in a little bit when I put it in formalin. But we can see the outlines of these hooks—the "thorns" that inspired the common name of thorny-headed worm—made clear by the chemicals needed to get this specimen into the balsam. When fully extended and pushed into a bird's intestinal wall, those thorns cut through tissues, helping to hold the worm in place. Traditionally, the number, size, shape, and arrangement of hooks are characteristic of a worm species, thus critical for identification of a specimen.

That same chemical treatment I used sixty years ago also lets us see muscle fibers in the body wall and the nerves that stimulated those fibers to make the worm move inside a bird's gut. This worm is a female; the space inside its body is filled with eggs. Its vaginal opening is plugged, revealing that it has already mated and at least some of those hundreds of eggs are fertile. The gods of chance delivered both a male and a female worm to a meadowlark on the Kansas prairies. The vaginal plug ensures that if those same gods of chance deliver another male thorny-headed worm, that male will have no chance to reproduce unless there is another virgin female inside that meadowlark.

Had that paid collector not aimed his .410 out his truck window and pulled the trigger, some of these eggs now in a museum specimen would have been deposited on the Kansas prairies every time 2549–50 defecated. By an incalculably small chance, a grasshopper would have eaten one of those eggs, maybe several, depending on the day, time of day, and weather. The chances of that one grasshopper living for the month, not being eaten by some predator before the worm developed enough to be infective, cannot be calculated, although for a particular grasshopper species, given enough time, available scientific talent, money, and access to prairie, the chances of a randomly selected member of its species living that long could be estimated. So that one thorny-headed worm in a meadowlark's intestine is a miracle of sorts; the fact that there were two of these worms, with a vaginal plug in one of them, doubles the miracle.

That fact is also tangible evidence that a worm species is doing the minimum required to survive in central Kansas: ending up in a grasshopper (pure luck), that particular grasshopper living for a month before being eaten (pure luck), that one grasshopper ending up in a meadowlark's gut instead of another bird species' gut in which it might not develop (pure luck), ending up in the same meadowlark gut as a member of the opposite sex (pure luck), ending up with a member of the opposite sex mature enough to mate (pure luck), and mating (typical parasitic worm behavior, provided all the foregoing pure luck). Chances of 2549–50 being the very one shot by a hired collector? Indeterminable, although if you estimate how many meadowlarks there were in Barton County, Kansas and what fraction of them would be sitting on a fence post at midday, you could make an educated guess. Whatever that chance is, another *S. neglecta* will die from that shotgun blast in the next month, joining the ten already in the freezer, destined to be dismantled by a scientist looking for the lessons that can be learned from worms.

Thus, even while finishing my work on 2549–50 in the early fall of 1964, my mind is extending that experience, building on it to form a worldview in which this difference between the individual and the species, between the one and the kind, is central to every thought I will have for the rest of my life, and shaping my reaction to personal, professional, biological, and political events. My daily news feed focuses on kinds: immigrants, gay people, Black people, Democrats, and Republicans, to mention a few; but my daily life deals with individuals, people with whom I do business, buying goods and services, sharing conversations, and sending emails. Those individuals rarely, if ever, match the images portrayed by some who talk constantly about kinds, often in derogatory terms. Sometimes I wonder what might happen if I could collar some state legislator, sit that person down at a microscope, pull out a drawer of museum specimens, and ask: How do thorny-headed worms of a single species differ among themselves?

To answer that question, this elected official would spend hours, then days, sometimes turning into months, making measurements, counting hooks, drawing then comparing anatomical features, and following up with statistical analysis. In the end, that person would never again comfortably equate kinds with individuals or vice versa, no matter what legislation was being proposed. I would then ask: How do *different species* of thorny-headed worms differ structurally, functionally, and in terms of the bird species' guts that will support their development into grown-up, mating individuals? The hope would be that after more months at the microscope, this lesson about the distinction between individuals and kinds would somehow permeate that lawmaker's every thought and action, to the point that it became a regular feature of their public persona. Never again would this person use those words so common in our current political discourse—immigrants, LGBTQ+, Blacks—in the same way as do their less-educated colleagues who'd never spent that time at a microscope. It's easy to label kinds; we do it every day. It's not so easy to label individuals. When we are introduced and learn about their lives, they become real people with parents, children, spouses, jobs, feelings, and responsibilities, just like the rest of us.

Those questions about thorny-headed worms are easy to answer if you have enough specimens, a nice microscope, and the right literature. The larger question, however, is of a different magnitude altogether: Where did thorny-headed worms come from? We know that these worms share some traits with other animals; for example, neither thorny-headed worms nor tapeworms have digestive tracts—a mouth, an anus, and a tube connecting them. But does similarity in one sense mean a common ancestry, and conversely, to what extent are differences real or superficial? This last question is of serious interest to evolutionary biologists, of course, but those same scientists apply their findings to humans and tell us that yes, most of what we consider differences among ourselves are superficial, and only in our minds have we made some minor genetic variations a serious problem for some of us. We are far more alike at the molecular level than some of us are willing to admit.

But we're not worms with a several-hundred-million-year evolutionary history. Had I known or been told at the time what molecular biologists would discover about thorny-headed worms half a century into the future, I would never have believed it. Nor would any of the scientists familiar with thorny-headed worms have believed it. As a group, these worms' closest relatives are microscopic animals just like those swimming around in my birdbath, animals with a telescoping body and a front end, covered with cilia, that creates a swirling water current drawing in bacteria and algae that then get crushed by a constantly chomping set of plates.[5] These free-living animals are called rotifers, a name derived from how those beating cilia appear under a microscope—like two wheels, spinning and sucking in food.

Back when I was dismantling 2549–50, anyone who claimed that thorny-headed worms were most closely related to rotifers would have been considered downright crazy because of the vast structural difference between the two groups, as well as their completely different lives—obligate parasites, in the case of thorny-headed worms, and free-living in the case of rotifers. Also, nothing in the structure or the life cycles of these groups even remotely suggests a relationship.

But the same molecular technology now used to solve crimes tells us that this anyone was not crazy at all, although they didn't know at the time that they were sane. Nowadays, we learn this lesson about the power of technology to alter the state of our minds almost every day; only parasitologists, however, seem to learn it best from how we treat our worms.

How and why that evolutionary divergence between rotifers and thorny-headed worms happened is a mystery that cannot be solved. Better to focus our attention on something that can be explained, such as how 2549–50 picked up the malarial parasites inside its red blood cells, now dried and stained, on that glass slide made weeks ago by the collector with his cigarette and shotgun. And also better to remember the truly important lesson about tangible evidence for what the world is really like, deposited in a museum, as well as the larger lesson about the value of museums as repositories for facts that cannot be denied by social media wizards. But here is a warning: Be careful if you decide to study malaria, mosquitoes, and thorny-headed worms; you'll end up fascinated by host-parasite interactions on a global scale. I guess the bigger lesson is to be careful if you give your kid a microscope. While you're still screaming about the lice they've brought home from school in their hair, they'll want one louse as a specimen, irrefutable evidence that they were infected by lice at a particular place in the world and on a particular day and that the stories they tell about those lice years hence are indeed true.

CHAPTER FOUR

FAVORITE MAGGOTS

Adapt or Die

The worms crawl in, the worms crawl out,
The worms play pinochle on your snout.

From "The Hearse Song"; traditional[1]

The first rule of fine dining is to never invite a parasitologist to dinner. But if you ignore this rule and do it anyway, you'll end up having unforgettable conversations about such topics as the beauty of shark tapeworms, puppies getting roundworms from their mother's milk, parasites that cause mice to lose their fear of cats, and the dangers of raccoon manure. No matter how much time and effort you've spent on the preparation and presentation of this dinner, someone is likely to list the worm species potentially acquired by eating it. In turn, however, you might explain why the Congo floor maggot is one of your favorites of all those flies that cause myiasis, an infection by maggots, second only to the ones that lay their eggs on the backs of small frogs, hatching into larvae that start eating the frog alive, reducing it to skeletal pieces in a couple of days. Then smile and ask everyone else about their favorite maggots and what kind of damage they can do. It's okay if a couple of your guests decide to go home early.

Scott Gardner, curator of parasitology at the University of Nebraska State Museum, once mentioned that the Congo floor maggot, distributed throughout sub-Saharan Africa, was his favorite. At the time, I thought maybe he just liked the sound of the name, somehow more melodious, or more ominous, than "screwworm," "blowfly," or "ox warble," all of which refer to flies that cause myiasis. Or maybe he liked the fact that Congo floor maggots are the only fly larvae known to feed on human blood. Adult flies lay their eggs on floor mats or dirt; the resulting larvae come out at night, pierce the skin with mouth hooks, and spend the next fifteen or twenty minutes sucking blood.[2] We assume that a victim remains asleep. Defense against attack is a bed that keeps you off the floor.

These maggots teach us that where you are and what you do determine how vulnerable you are to things that can get in or on you. Most of us in the so-called developed world know this fact but tend to ignore it unless we decide to go traveling. That lesson also applies, of course, to situations far removed from sub-Saharan Africa, including parasitic ideas that can get into your brain if you get too close to their source, for example politicians spouting fear and loathing of LGBTQ+ people, some of whom could easily be your personal friends or relatives. Those parasitic ideas, however, share a feature with fly larvae that can suck your blood, and that is the fear they can produce when you're exposed to them, whether that fear is warranted or not.

Compared to the Congo floor maggot, my favorite—larva of a fly named *Megaselia scalaris*, or scuttle fly—seems almost benign, at least to most people, although it's been known to invade human intestines or urinary tracts, thrive in corpses where it functions as forensic evidence for time of death, and destroy beehives.[3] You've no doubt seen *M. scalaris* in garbage cans, dumpsters, public washrooms, or on bananas you've brought to a picnic and just shooed it away, not knowing about its invasive behavior. But the scuttle fly, found all around the world, also has its own parasites, single-celled ones with flagella, which is why it is my favorite and why for the past fifty years, I've worn a silver bracelet with images of these parasites to remind me of what they and their hosts taught me about survival in changing environments. Those parasites turned out to be masters of adaptation, and if there is anything we humans need to learn now, it's how to adapt to change, especially on a rapidly warming planet swarming with people trying to find a better life.

I came to truly appreciate the role that flies play in human health when I was doing a postdoctoral fellowship at Rutgers, in New Brunswick, New Jersey, studying the biology of *Leishmania* species, parasites that can cause death or horribly disfiguring infections in humans. These parasites can also infect other animals, including dogs, cats, rodents, and reptiles, with dogs and rodents being notorious reservoirs, especially throughout the Middle East. *Leishmania* parasites are carried by tiny flies called sand flies, of the genera *Phlebotomus* and *Lutzomyia*, with different species of these vectors in the Old World and the New World. The mysterious but prevalent diseases they carry are known by various names, such as white leprosy, espundia, Delhi boil, Oriental sore, dum-dum fever, and kala-azar.[4] A Google search on any of those terms produces a nice introduction to not-so-nice tropical diseases.

Leishmania parasites are flagellated cells, or protists, members of the family Trypanosomatidae, commonly called "trypanosomatids."

They evidently have been with us at least since our smallest mammalian ancestors evolved and were dodging dinosaurs, as a species named *Paleoleishmania proturus* that was found in fossilized sand flies stuck in early Cretaceous amber almost 150 million years ago.[5] As old as these organisms are, it was only in 1901 that they were recognized as parasites, by William Boog Leishman and Charles Donovan, when Leishman examined spleen tissue from a soldier who had died in Dum Dum, West Bengal, India. Two years later, Ronald Ross, winner of the Nobel Prize for discovery of the malarial parasite life cycle, named the microscopic bodies found in that spleen as the species *Leishmania donovani*, a name assembled in honor of those two fellow scientists. Discovery of new *Leishmania* species, their diverse clinical manifestations, reservoir hosts, and vectors has continued ever since, as has the molecular biologists' explorations of their evolutionary origins and relationships.

When I came to the University of Nebraska, I brought part of my postdoc research program with me and spent the next fifteen years as a cell biologist studying these trypanosomatid flagellates. I ended up with about twenty different species and strains of these protists in culture, including several that had originally been isolated from insects and were considered noninfective for people and three that had been isolated from Old World lizards. These last three led to the choice of larval *M. scalaris* as my favorite maggot.

This research was the easiest I've ever done because the organisms we used grew well in culture media containing rabbit blood, thus providing lots of material and allowing me to very easily plan my research around teaching obligations. I spent weekends preparing for classes, did research on Tuesday, bled rabbits on Wednesday, made culture medium on Thursday, transferred cultures on Friday, then went to the golf course and drank some beer. Because some of our organisms caused human disease, it was easy to get grant money and attract students who wanted to work in the lab. Also, there were virtually no animal care and handling regulations at the time. This research could not be done at most American universities today for several reasons, the main ones being the burden of regulations and facilities required to do any kind of work with live vertebrate animals and pathogenic organisms. To my knowledge, none of us got infected; nor did we start any fires in the lab with all the alcohol we spread around to keep things sterile. That lab bench wipe-down and hand-washing was an early version of the COVID-19 cleaning behaviors that characterized local restaurants in 2020 and continues due to the SARS-CoV-2 virus's evolution and politicians' ignorance.

Among the graduate students who came to our lab was a young man named Norman Dollahon, who quickly became "Norm" in all our lab talk. As we discussed his possible dissertation projects, the question came up of whether *Leishmania* species that infected Old World lizards would be infective for New World lizards. This question was actually a biogeographical one about the global distribution of both hosts and parasites and especially the constraints, ranging from continental drift to ecological factors or presence of vectors, that might operate to determine those distributions—a lizard version of a typical human disease problem. We had three *Leishmania* species in culture that had originally been isolated from Old World lizards—*L. agamae*, *L. adleri*, and *L. tarantolae*—and we could get a supply of New World lizards that we either caught locally, bought from pet supply sources, or had shipped to us from various places by colleagues. Although none of these parasites were believed to be infective for humans, even this research involving lizards could not likely be done today because of the animal care and use regulations now in force throughout the nation's academic institutions.

But back in the 1970s, nobody seemed to care that we collected lizards and brought them into the lab. So Norm started his research, aiming to discover whether parasites from one continent could infect denizens of another continent across the ocean. The parasites we had in culture were members of the parasite family Trypanosomatidae. This work was of interest because lizards can't fly across great distances as do the migratory birds described in chapter 2. Because his experimental hosts were of different species than those from which his parasites had been isolated, this project was a version of the cross-infection cases so familiar to virologists. Since SARS-CoV-2 has spread throughout the world, we've become quite familiar with the movement of infectious agents from one species to another, from bats to people, for example, and also to our pets and local wildlife. But lizards don't have and have never had airplanes to help keep them in regular contact with relatives across the ocean; that's why Norm's research was a study in evolution and biogeography—the factors that determine distribution of life on Earth. Although the involvement of *M. scalaris* was not planned as part of his work, it became a most intriguing contributor to the research in our lab, in the process earning its parasites a place on my silver bracelet.

There was a guinea pig colony in our building, and because of the way these animals were kept, over pans that collected urine, feces, and scattered food pellets, that colony was a great source of *M. scalaris*. Their larvae would live in any kind of organic waste, so they were also developing throughout our building anywhere there was garbage or food left around by students and faculty members. They were thus behaving in our building the same way they behaved around the world. Nobody knew where they came from originally, nor did we truly understand at the time how aggressive they could be, but we put out flypaper in an attempt, unsuccessful it turned out, to control their population in our lab. But in the animal rooms, the pans below the guinea pig cages were seething with maggots. When you can thrive in a mixture of urine, feces, and garbage, you have a certain kind of power to survive under circumstances most folks believe are pretty terrible. That power to thrive in waste, showing me that garbage can hide an opportunity, is the source of my love for larval *M. scalaris*.

Norm's research involved experimental infections of New World lizards with those species of *Leishmania* originally isolated from Old World lizards.[6] Because all his lizards were captured in the wild, he had to follow the same general protocol we used with any supply of wild-caught hosts used in infection experiments: dissect a third of them at time zero, the start of your experiment, to determine whether they are naturally infected; then infect a third of them; and finally, keep the last third for dissection at the end to determine whether your host supply stays uninfected. By following this protocol, you end up with a time zero control group, a time "t" experimental group, and a time "t" control group. If both of your control groups remain uninfected but your experimental group has parasites, your experiment is valid; it's also valid if your experimentals are parasite-free, but that just means you have another question to answer. The time "t" in this case was about three weeks. He therefore had to keep his lizards alive for that period, which meant feeding them mealworms and crickets and providing them with water and heat. You can see where this is headed: mealworms and crickets are cultured in organic media such as rabbit food and wheat bran, and anything that ended up wet and organic would attract *M. scalaris*. And of course, it did.

Norm used the following lizard species in various aspects of his research: green anole, lesser earless lizard, brown basilisk, Mediterranean house gecko, yellow-headed gecko, desert iguana, six-lined racerunner, four-lined ameiva, southern alligator lizard, and collared lizard. Except for the Mediterranean house gecko, an import gone wild, these species are all restricted to the New World. To infect these lizards, he injected cultured flagellates into the peritoneal cavity, that space surrounding the intestines. To determine whether the lizards were infected, he had to do cultures of blood, liver, spleen, and leg muscle. He dissected the animals under sterile conditions, homogenized the tissues, put a sample of the homogenate into blood-based culture medium, and checked that culture microscopically over the next week or two. If flagellated cells—those with a long, whip-like flagellum extending outside the cell—developed in the cultures, the animal had been infected. If the control animals were not infected, Norm was looking at a discovery of interest to parasitologists in general and a publication that would advance his career. If the parasites were the same kind he injected, the cultured cells would look something like this:

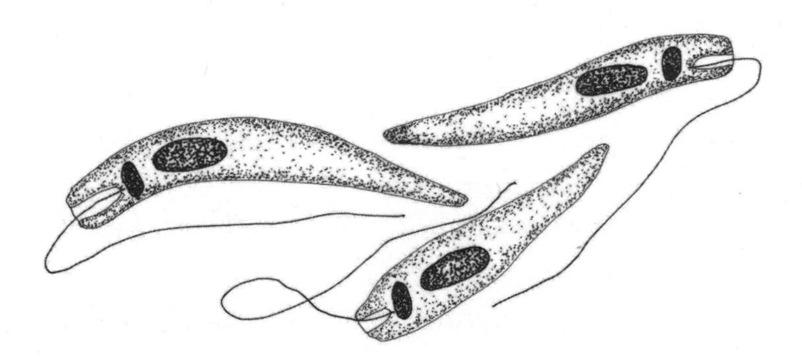

The larger, oval-shaped structures in these cells are the nuclei, containing DNA, but the smaller, oval structures at the base of the flagella also contain DNA and are called kinetoplasts, which are always associated with the flagellar base.

Among the lizards sent to me from Costa Rica was a large, beautiful, plumed basilisk about eighteen inches long. The basilisks from Costa Rica were great experimental animals; they'd eat mealworms and crickets and bite the hell out of you if you weren't careful handling them. Because it was such a magnificent specimen, I wouldn't let Norm use the plumed basilisk in his research, so we kept it in the lab as a pet for a few weeks until one day it was very lethargic, and a day later, it died. But we were curious about the cause of death, so he dissected it to check for parasites using his sterile culture techniques.

A naturally infected New World lizard would be a publishable research note and probably a new species of *Leishmania*, a real plus for a grad student anticipating a future job search. We also noticed that the plumed basilisk's skin had some small holes in it but passed that observation off as an unsolvable mystery. It had been housed in a large aquarium, provided plenty of insect larvae for food and a lighted bulb for warmth, and its water dish, into which it had kicked a bunch of bran and other stuff, was crawling with small maggots. I didn't think much about those maggots, except that I suspected they were larval *M. scalaris*. Adult *M. scalaris*, family Phoridae, are about the size of *Drosophila* (fruit flies) but are much more aggressive and tough. And they were everywhere in our building, a real nuisance. A few days after this lizard funeral took place, Norm came into the lab.

"That muscle culture was positive," he said. "But blood, liver, and spleen were negative."

Wait a minute, I thought; a *Leishmania* infection should be in the liver and spleen, but in the leg muscle?

"And they're not *Leishmania*," he added. The flagellated parasites in his culture didn't have the correct structure to be a *Leishmania* species. Cells of the parasite family Trypanosomatidae have a nucleus, a flagellum, and that DNA-containing structure called a kinetoplast, always associated with the base of their flagellum. If these parasites had been *Leishmania* species, these structures would all have been at the front of the cells, as in the previous picture, but in Norm's stained culture smears, some of the flagellates from this lizard culture had kinetoplasts that were either beside the nucleus or even in the posterior part of the cell, as in the following illustration, but with no undulating membrane, the latter a distinguishing structural feature of related species that cause serious disease in humans and domestic livestock.

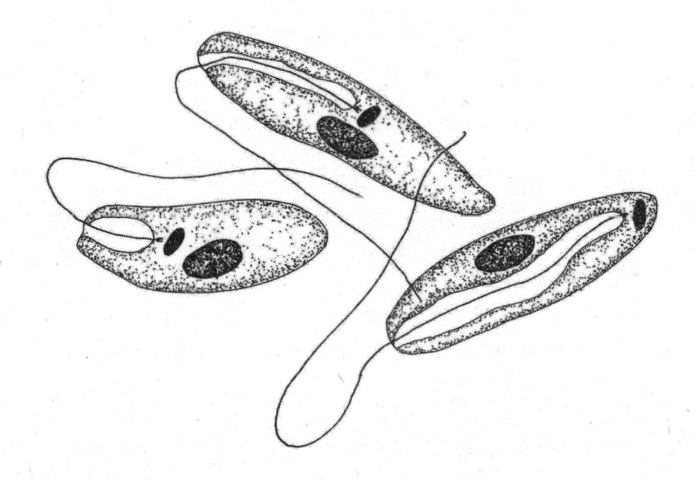

These parasites were obviously something new, at least to us, and because they grew so well in culture, we naturally kept them alive for research by weekly transferring some of them into screw-cap tubes with two milliliters of blood/citrate culture medium. Two questions routinely came up in our lab discussions: What should we do with these cultures? That question was easily answered: Keep them alive until you figure out something interesting. And, of course, the most pervasive question in biology: What is it?

Among the undergraduate students who'd walked into my office in the late 1960s just looking for something interesting to do was a young man from western Nebraska, Pierre-Marc Daggett. He was brilliant, creative, and completely uninhibited, and he spent quite a bit of time in the library reading ancient papers and talking about infections that seemed almost science fiction to the rest of us. Naturally, he gravitated to the cultures Norm had isolated from that pet lizard. It didn't take Pierre long to decide, based on the structural changes that took place as a culture aged and exponential reproduction slowed, that he was dealing with a species of the genus *Herpetomonas*, similar to parasites in houseflies and characterized by having the kinetoplast at the posterior end of the cell but with no undulating membrane. We encouraged Pierre to describe the one we had as a new species; he didn't need much encouragement and went right after the task.

How do you describe a new species in a way that it gets published, thereby adding your name to the literature of biology and this newly discovered species to our inventory of life on Earth? The answer to this question varies greatly, depending on what kind of species has been discovered and when, but anatomical measurements and illustrations have routinely been required since 1758, when Linnaeus established our rules for naming organisms with the tenth edition of his classic work, *Systema Naturae*. In the case of trypanosomatids, when Pierre did this work in the 1970s, he needed to provide not only measurements showing his parasites were structurally different from related species but also observations about the kinds of culture medium used and changes in form as the flagellates multiplied over time. And he had to deposit a living culture in the American Type Culture Collection, a repository of living material available to other scientists.[7]

While doing this work, Pierre discovered that this parasite was much more tolerant of its growth conditions than many other related trypanosomatids, which can be notoriously fastidious in culture, if they can be cultured at all. He also discovered that the pans below the guinea pig cages were crawling with *M. scalaris* maggots and that they were all infected, as were the adult flies—a perfect storm of parasite transmission subsequently carried throughout our building, not unlike some super-spreader political event in the age of COVID. He proposed a name for his new species—*Herpetomonas megaseliae*—based on its structure and its source, in essence honoring the fly species that carried it.

By this time, we'd decided that Norm's leg muscle cultures from the plumed basilisk were positive because he'd ground up infected *M. scalaris* maggots that had been eaten by the lizard, burrowed through the lizard's stomach or intestinal wall, and buried into the leg muscle, carrying their parasites with them. But we didn't rule out the possibility that the maggots had burrowed into the lizard from the outside, because of those small holes in the skin. The flies were doing something in our lab and throughout our building that they'd likely done for millennia on corpses in various battlefields and in garbage dumps around the world.

By this time, Pierre had also discovered that this flagellate differentiated on schedule. In other words, once they stopped multiplying, the cells changed structurally, their kinetoplasts migrating posteriorly to the end of the cell. Predictable changes in cell structure, especially in an easily cultured line of cells, are a gift to researchers because cellular differentiation is the process by which complex organisms, such as ourselves, are built and by which cells attain their final functions. But kinetoplast migration, as was happening with *H. megaseliae*, is also a developmental feature of related parasite species that cause human and veterinary disease, so this sequence of changes was a real discovery.[8] We immediately began thinking of *H. megaseliae* as a model for studying this differentiation process also found in important parasites, such as those that cause African sleeping sickness, but with an easily managed and cooperative species available from the fly pests in our building.

Megaselia scalaris and its parasites had therefore delivered a system for understanding whether change produces virulence, and if so, how. Although the system came packaged in a microscopic cell, the general concept it demonstrated was clearly applicable to situations far beyond a laboratory in Nebraska: structural change is related to, an indicator of, or a predictor of functional change. We see that principle manifested in every new automobile model and every "improvement" in smartphones. We hear it promised with every election cycle. Historians remind us that certain kinds of changes—extreme nationalism, growing racism, authoritarianism—can let us predict social pathology, with Nazi Germany being a prime example. *Herpetomonas megaseliae* is just a fly's parasite, but structure-function relationships are characteristic of all living organisms and all the things they build, including corporations and governments. So when it became obvious that *H. megaseliae* differentiated on schedule, we jumped on that observation.

Pierre Daggett took the lead in this research, eventually writing his MS thesis on the biology of this flagellate and for the next three years using it as one of the parasite species involved in one of the most remarkable doctoral dissertations I've ever read: a study in which he assessed the ability of different trypanosomatid species that were originally isolated from insects to survive in mammalian cells and mammalian hosts and extend their survival times through an adaptation protocol. This research addressed an important evolutionary question experimentally, namely how parasitic relationships come to be established and how organisms evolve into parasites occupying certain hosts.

The question of how and why parasitism arises is one that simply cannot be answered for certain because by the time we humans observe a parasitic relationship, all the evolutionary changes required to establish it have already been accomplished, probably millions of years earlier. But with the right organisms in the lab, we can mimic a proposed scenario in which a species acquires the ability to survive within a potential host's body. Pierre's research did just that, in the process demonstrating that some flagellate species can increase their survival in mice but others could not. The ones that could increase their survival time released a new set of molecules into their culture medium after being exposed to mice, molecules that could potentially inhibit a mouse cell's ability to digest an invader. I have yet to read another dissertation so characterized by brazenness, audacity, and an inhuman amount of work. It was a lesson in what it takes to make a conceptual contribution to a nagging issue in evolutionary biology and to do it experimentally.[9]

While all this culturing and describing was taking place, however, Norm Dollahon was also using *M. scalaris* maggots in some of his research, with a special focus on what happens to the flagellates when a lizard eats their maggot host. He was actually trying to determine the process by which a basilisk's muscles ended up with parasites that should have been in the liver instead. Keep in mind that his research was, like Pierre's in many ways, far outside the realm of parasite biology as it was practiced globally back then. Their published papers raised the question of whether these parasites were behaving as expected, at least in the lab, and by inference in nature. I love it when students demonstrate parasite features that challenge our perceptions, knowing that such demonstrations make us challenge perceptions in realms far distant from parasitology and draw analogies that apply to the daily news.

Pierre Daggett's new species description was published in 1972, and as a result, we were able to publish a number of other papers on the biology of this flagellated parasite living inside *M. scalaris*. For a variety of reasons, it's important to be able to put a scientific name on organisms being studied, and nowadays it's probably impossible to get experimental results published in a peer-reviewed journal without not only a scientific name but also collection numbers for voucher specimens deposited in a reputable museum and DNA sequences posted in an internet site named GenBank. Norm did manage to get a paper published without the scientific name, however, by feeding *M. scalaris* maggots to lizards and then isolating the trypanosomatid parasites from both the intestinal contents and feces of these lizards.[10] Evidently the anonymous reviewers of his submitted manuscript felt the survival of these parasites under such conditions was too important not to publish, regardless of their namelessness.

With such results, Norm showed that *H. megaseliae* was doing something it was not supposed to do, something that its relatives, those in the same family, would never be able to accomplish, namely survive transition through a vertebrate animal's intestine. Larval *M. scalaris* were earning their place as my favorite maggot and on my bracelet by delivering a gift: a parasite that was far more flexible, opportunistic, and daunting than its relatives. This flagellate was essentially saying to its relatives, most of whom are considered quite important by humans: "I can and will do things that you cannot or will not do."

In essence, this parasite was teaching us things that were unexpected, things that completely altered our sense of how parasites operated in nature, surviving at least for a while under conditions where it should have perished, like inside a lizard's intestine or a mouse's liver. In doing so, it was also teaching us how to think about parasitism, the most common way of life among Earth's animals, and especially about its "more important" trypanosomatid relatives, that is, the ones causing diseases, sometimes fatal, in humans and livestock. Those lessons shaped my approach to science for the rest of my career, mainly by leading me to always listen to what some parasite was telling me about its life, telling me to get my questions from observations of nature instead of from the published literature or my colleagues. That approach seemed to be infective also, especially a decade after *M. scalaris* came into my life and the University of Nebraska opened its Cedar Point Biological Station in the western sandhills, a parasitological wonderland about which we knew little or nothing but had plenty of opportunities to explore.

Then one day in September 1978, I was walking through a local shopping mall where artists were displaying their work and happened to stop at a table where a woman named Jackie Lusher was making jewelry. On a whim, I asked her if she could make me a silver bracelet with figures of *H. megaseliae*, the parasite Pierre had described from my favorite maggot, and drew the pictures for her on a small piece of paper. She smiled and said, "Sure." A week later, I stopped by her house, paid her, and slipped that bracelet on my right wrist. It's been more than forty years since I first put on that bracelet, and I still wear it all the time, carrying its messages: *I can and will do things you cannot; I'm willing and able to survive under conditions that you cannot; and the world of infectious organisms is quite different from what you believe it to be.* I wear a parasitology lesson on my wrist, one that's also a life lesson taught by a microscopic flagellate living in my favorite maggot.

Decades after all these events happened, the feel of that silver band on my wrist brings to mind the dreams, struggles, and successes of young people trying to satisfy their curiosity about how one species survives inside another. My doctoral adviser at the University of Oklahoma, John Teague Self, told me repeatedly that the greatest reward of an academic life would be the success of my students. He was correct; *H. megaseliae* repeats that reminder every time I put that bracelet on my wrist. But the bracelet is also a reminder that a careful and unbiased study of nature leads to explanations, ideas, and concepts that are far closer to reality than the blathering of pathologically insecure and uneducated men in positions of power. In turn, that unbiased study teaches us that we humans have the capacity to understand what is truly happening around us, no matter how much we want to believe something different, and that flexibility, along with a willingness to adapt to changing conditions, is how to survive in challenging times.

There was no internet when these students were discovering just how adaptive was this parasite of my favorite maggot. Had all this work been done the past few years, those Friday afternoon conversations would have quickly moved beyond cell biology and into the realm of public discourse about all kinds of global issues. Foremost among those issues would be climate change and the daily news as prelude to environmental disaster. Eventually someone would make a smart-aleck comment, something to the effect that if you want to survive in a postapocalyptic world, it might be wise to take lessons from a fly that thrives in guinea pig urine and feces and its flagellated parasite that can do a bunch of things its relatives cannot. And because this conversation is only bar talk, it's okay for one of us to substitute "will not" for "cannot," humanizing the model delivered by a tiny fly and its parasites.

CHAPTER FIVE

THE PLAINS KILLIFISH

Distant Events and the Nature of Connections

> It will come about that every living creature which swarms in every place where the river goes, will live. And there will be very many fish, for these waters go there and the others become fresh; so everything will live where the river goes.
>
> Ezekiel 47:9

One day during the summer of 1975, two young men—Mike McCarty and Rick Goble—went to the South Platte River south of Brule, Nebraska, cast their nets in the water, caught some small fish, and came back to where they were staying at the time and where I was teaching a summer class, a university field camp named Cedar Point Biological Station. Mike and Rick examined their catch, including a small striped fish named the plains killifish, then came to me with a question: What are these white things on the gills? I answered that they looked like parasite cysts and then asked them to pose with this fish for a photograph.[1] At the time, that was a pure guess but an educated one, a teacher's defense against looking stupid.

That answer, however, and my subsequent efforts to understand what those white cysts really were and how they got there inspired nearly thirty years of research on the denizens of the South Platte River and their community of parasites. It took most of those years before I woke up to the lessons that the plains killifish, *Fundulus zebrinus*, and its parasites were really teaching me about the impact that distant events had on my everyday life and on the lives of people who were paying my salary at the time. That's what sometimes happens to scientists: it takes them a while to figure out the larger lessons they've learned from their research and why those lessons are more important than the research itself.

Parasites are notorious for their complex but fixed or evolved developmental cycles, ways of life that are as nonnegotiable as political platforms and religious beliefs. But the South Platte River is equally notorious for its fluctuations, rather like the American stock market but far more dramatic. Streamflow can vary significantly from year to year, month to month, and sometimes, it seems, even week to week.[2] If you're a parasitic worm in this kind of a river, you either find hosts in which to complete your life cycle or you die. That burden is the product of evolutionary history, especially challenging if there are several different hosts required, in sequence, and each of those has its own relationship to the river. The plains killifish has at least seven different kinds of parasites that depend on it. Some of those parasites are single-celled ones that live on the gills; others produce those cysts discovered by Mike McCarty and Rick Goble; still others, worms with big hooks on their rear ends, attach to the fins; and one burrows into the eyes, body cavity, and internal organs, especially ovaries. And like their host fish, these parasites' survival requires movement through a physical environment that is sometimes calm but other times violent.

What produces this highly variable environment in which fixed systems of survival operate? The answer is somewhat distant events, namely annual differences in Rocky Mountain snowpack, the same events that shape grain production across much of the American Great Plains, with subsequent fluctuations in the commodities markets and the price of bread in your local supermarket.[3] The events that influence South Platte River streamflow are thus distant not only in space—several hundred miles across the prairie from where Mike and Rick collected their plains killifish specimens—but also in time. Snow that falls in January ends up irrigating corn in July. And depending on the parasite species involved, snow that falls in January also ends up dictating how many worms are in a killifish in July; if it could talk, a fluke larva that must penetrate a fish to survive would have a nice conversation with local farmers about South Platte River streamflow and its impact on their lives.

What produces variations in Rocky Mountain snowpack and thus variations in streamflow and parasite numbers in a Great Plains river? No one really knows, but the answer probably involves something that happens over the Pacific Ocean. Some distant chain of events produces moisture that falls to Earth as snow in Colorado and Wyoming. The amount of water moving down the river toward the Gulf of Mexico every summer is directly related to the amount of snow in these mountains. One of the parasites that must negotiate this aquatic environment is the larval worm that lives both inside the killifish's eyes and inside its body cavity. In drought years, a female's ovaries could be mostly worm tissue; in those same years, there could be fifty of those worms in a single eye, packed in behind the retina. In flood years, there are many fewer of these worms, and fewer fish are infected, although because fish can live a couple of years, the rivers' behavior in one year may not influence parasite numbers until the following year.[4] That's another lesson from parasites: although the effects of those distant events may not be immediate, they may linger.

If the South Platte River had flowed at a constant rate during my decades of counting parasites, that lesson about the impact of distant events on local ones would never have been learned. The numbers of different parasites in these fishes would not have varied so dramatically. But the river's behavior revealed how some kinds of animals avoid that impact because of their inherited or evolved traits, but others cannot. Even for parasitic worms, the circumstances and places of their birth as well as their inherited traits dictate not only how they negotiate the good times and bad but also how they survive. That's a life lesson you can supposedly learn from a textbook or from your parents, but it never sticks in your mind the way it does when you experience it by going to a braided prairie river, netting fish, dissecting them, and counting their parasites over an extended period.

Throughout those thirty years since taking the photograph of Mike McCarty, Rick Goble, and their catch, I personally collected well over a thousand plains killifish, and my students collected another several hundred of not only *F. zebrinus* but also various other species of small fish. This parasite hunt involved a whole lot of time wading in the river with a seine and a lot of additional time, often through the night, at a lab bench and microscope. What did we actually do during these decades? The answer to that question is probably best illustrated by a quote from the scientific literature, a paper I cowrote with two of my graduate students in the late 1990s that was published in the *Journal of Parasitology*:

> These collection conditions satisfy the homogeneity requirements for parasite field data sets for each sample (Janovy et

al., 1992). Cohorts are obviously lumped in such collections, but seine mesh size selects fish mostly 3.5 cm and larger, i.e., older young of the year (6–8 wk post hatching), second-, and third-year fish. Samples consisted primarily of second-year fish as estimated by size (mean total length = 6.21 cm, variance = 0.95 cm, n = 1,219; males = 6.32, var = 0.97, n = 595; females = 6.10, var = 0.91, n = 624; cf. Minckley and Klaassen, 1969; Brown, 1986).[5]

All that experience is summarized in those hundred words; anyone who's done similar field and lab work could read those words and easily reconstruct, in their minds, our days on the river and our nights at a microscope. The names and dates in parentheses are citations of references listed in the bibliography, so in telling what we did and how we did it, we referred to three other publications, and thus the research of at least three other people, to validate our actions. That is typical behavior in the sciences; we justify our claim for the numbers, that is, our observations—how meaningful they are and how we've interpreted them—based on what others have shown to be useful approaches in similar studies. Not all readers of this particular paper would agree with that claim, but enough did so that our numbers were published, along with our interpretation of them. In this manner, the work of those earlier scientists was also validated; that's why we routinely look through the published literature while designing and doing research.

Why did we collect and cut up 1,219 fish? One reason was to satisfy the requirements of various statistical tests applied to the observations; that's the reason used at the time this work was done. But the real answer is a relatively simple one, although it took those decades to demonstrate it. Parasite success, measured in terms of numbers, had to be assessed through good times and bad before we could draw conclusions about the role that environment played in controlling a population. This time requirement is basically no different from that needed to determine whether, for example, a business, a marriage, or a political campaign will succeed or fail; only the length of that time varies between a scientific research project and a political campaign or a piece of legislation. In other words, it was only in retrospect that we truly understood what we'd done and that nature had taught us a powerful and unforgettable lesson about the impact of an environment on those who live within it. The river dictates how some parasites thrive, that is, how heavily infected are the hosts, but other parasites seem to ignore the river's orneriest behavior. It doesn't take much casual conversation among parasitologists before this lesson ends up being applied to ideas, beliefs, and practices by various people and agencies, with the South Platte River replaced by that gushing river of information sweeping over us continuously.

The fact that this environment was a river translates easily into a metaphor. We are surrounded by this flow of people, ideas, money, news, pictures, words, and decisions that define our existence. None of it gets repeated exactly, and a lot of it, like Rocky Mountain snowpack, originates halfway around the world. Some of that information infects our minds and makes us behave accordingly. The Chinese social media platform TikTok is an excellent example of this principle at work. Launched in 2016 and hosting short videos made with web cams and smartphones, at the time this book was written, the TikTok app had 2.7 billion downloads and a billion daily users—about 30 percent and 12 percent of the human population, respectively—and its parent company, ByteDance, had offices in at least seven different countries.[6]

What makes TikTok so infective? Its sharing functions, applied to both creation and viewing of videos, is one of the main factors; when two or more people are involved in the creation of a product, they are invested in the results, a form of intellectual and emotional infection. TikTok users are essentially free from any constraints on their creative impulses, easily linking those impulses not only to commonly available and extremely powerful technology such as iPhones but also to like-minded users around the world. And much of TikTok content involves stories—short ones, yes, but storytelling has been an effective transmission behavior throughout human history and across cultures. Finally, a whole lot of TikTok videos are downright entertaining, and humans, especially young ones, are suckers for fun—they are not always immune to a good story, with a message, set to exciting music. A social media platform is thus producing its own river of information and influence, otherwise known as exposure and infection, with parasitic ideas differing in their success depending on the times, rather like worms in the South Platte River fishes.

Like humans in the age of TikTok, Instagram, and Facebook, along with the ideas they transmit, the fish is living in a constantly changing environment, and so are its parasites, and neither participant in this system can predict or control what will happen next. What we have, as a result of that question from Mike McCarty and Rick Goble at the beginning of this chapter—What are these things?—is a model for how species that are locked into certain life patterns deal with constant change. These fish and their parasites might as well be human beings, locked into certain socioeconomic conditions, certain behaviors, the circumstances of their birth, and political or religious beliefs they are too proud to reject or too ignorant to understand while suddenly dealing with a pandemic in which the infectious agents range from viruses to virulent language uttered by people who are not very smart or well educated, especially about disease transmission.

Not only do the parasites in and on this fish live in a constantly changing environment, but their developmental cycles are also evolved and thus fixed, resembling, in this regard, the job skills of many American workers, especially midcareer ones or those in a manufacturing industry. Labor statistics over the past few decades, for example, show that as the national work environment has changed in accordance with a growing service economy and technological advances in manufacturing, opportunities for those without appropriate skills or training, especially people in the twenty-one to fifty-five age range, have been greatly reduced. Electrification of automobiles is a prime example of this situation; electric vehicles have far fewer parts than fully mechanical ones and thus are easier to build. On a global scale, electric vehicles are here to stay, with sophisticated engineering approaches aimed at making them able to go increasing distances between charges. In the United States, across all industries, over five million manufacturing jobs were lost during the first decade of this century, the vast majority because of automation. Technology is thus functioning like a South Platte River flood, erasing opportunities for some workers to occupy a way of life while having little impact on others.[7]

Furthermore, there are four different kinds of these developmental cycles among parasite species that live in and on the plains killifish, as shown in the scene below. A heron (a) with an adult fluke in its gut sheds eggs in its feces (b). Those eggs hatch inside snails as first hosts in which infective larvae develop (c), then leave the snail and penetrate the fish. Worms that attach to the gills and fins (d and

e) produce free-swimming larvae that must encounter a fish. Tiny single-celled ciliates swarm over the gills and reproduce by simple division (f); and those cyst-producing parasites on the gills, the ones first seen by Mike McCarty and Rick Goble, must pass time in free-living, earthworm-like, aquatic worms before producing infective stages (g). These differences mean that there are also up to four different kinds of transmission mechanisms, two of which are influenced not only by the amount of water flowing down from Colorado mountains but also by the presence of other species of animals that both share the South Platte River and are themselves influenced by streamflow. For example, herons carry the adult stages of those worms in the fishes' eyes and ovaries, but other parasites must also negotiate river conditions to survive. The whole integrated system—fish, parasites, and river—looks something like this picture:

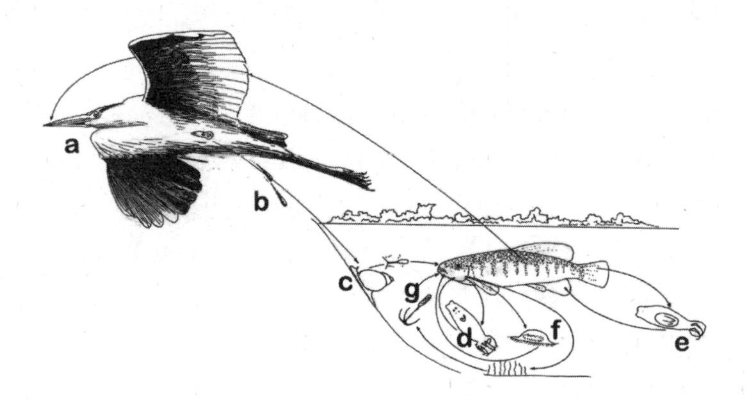

Thus, we have a truly complex set of relationships, so the specific question to be answered by digging through all these fish is how this system of interdependent relationships and stereotyped developmental events reacts to a constantly changing environment.

It's not very difficult to understand how this same question applies to human societies. I believe this interaction between inheritance and ecological conditions is why we should all become parasitologists, serious amateurs if not professionals, looking for infectious agents, identifying them, and trying to figure out how they evolved, how they are sustained, and how they manage to move through populations of hosts, including human minds. You don't need to go wading in the South Platte River to play this role. Your flow of information is as accessible as the television remote or your smartphone, and your laboratory is a local library. But you'll need to think like a parasitologist, asking how a parasitic idea evolved, what resources allow it to multiply, how it is sustained and transmitted, and whether, like some of the parasites in a plains killifish, it seems to affect the way its hosts view the world.

The worms that live in the fish's eyes are named *Posthodiplostomum minimum*; at least that's what they were called back in the 1980s and '90s when all this research was being done.[8] In conversation, we call it *P. minimum*, a spoken version of abbreviations typically used for scientific names once they've been used in a paragraph. In casual talk among biologists, rules are often suspended, especially rules for being careful with your interpretations. Thus, big meanings come from small talk: the manner in which Rocky Mountain snowpack influences the number of worms in a fish's eye five hundred miles downstream is actually a lesson in what kinds of events can happen, to use a recent example, when someone breaths in bat guano twelve thousand miles away, picks up a deadly virus from that exposure, then gets on an airplane to Seattle.

We don't really know what happens over the Pacific Ocean or when exactly it happens to produce moisture in the western United States, although El Niño and La Niña are the extreme situations, driven by unknown forces that influence ocean currents.[9] All we know is the result: in some years, a deep layer of winter snow blankets the Colorado Rockies; in other years, there's not so much. If modern science could predict these differences a year or two ahead, that prediction would be of substantial importance to American agriculture, especially grain production throughout the Great Plains. But that something and when it will occur are a mystery; we, Great Plains farmers and ranchers, *P. minimum*, and all the other parasites on *Fundulus zebrinus* live with the aftereffects. When the snowpack is heavy and deep and spring temperatures are just right, the South Platte River is bank-to-bank, swift and deep, washing away an essential resource for *P. minimum*, namely the snails that support its larval stages. When the snails are washed away, so are the larval stages. None are available to leave the snail host and penetrate a killifish eye, so none are available to mature into adult worms when a heron eats a fish. The whole system collapses in that year because of whatever the Pacific Ocean does to influence Rocky Mountain snow.

If you're the fish, it's probably okay that this *P. minimum* system collapses. But if you're a parasitologist, the collapse of some complex system due to a distant event over which you have no control is a lesson easily extrapolated to many other facets of daily life, such as health care in the United States during a pandemic. When bodies of COVID-19 victims must be stored in refrigerated trucks, critically important equipment such as ventilators are in short supply, there are thousands of deaths beyond those expected in nonpandemic times, and politicians fight against vaccine and mask mandates, you know that we are victims of infection events that took place half a world away but eventually ended up on our shores, having arrived on the wind, albeit carried in a metal tube with wings.[10]

Parasitologists, accustomed to such complexity as wrought by the evolution of multistage life cycles, have no problem drawing this analogy between different and unrelated fixed systems—worms in a fish's eye and a nation's healthcare establishment—each subjected to unexpected and uncontrollable change. From thirty years of research, we know that eventually the river will calm down, snail populations will explode, and *P. minimum* will thrive. We're also confident, from knowing about past pandemics, that whatever one is currently a problem is neither the first nor the last of its kind. History shows us that whatever lessons we're learning about and from worms in a fish's eye are applicable to our life beyond the South Platte River banks. The real question, of course, is whether we'll respond to that next pandemic better than the fish and worms respond to their respective environmental fluctuations.

Conversely, however, other parasites are far less affected by South Platte River streamflow than is *P. minimum*. One of the relatively unaffected parasite species is a microscopic but beautiful worm that lives attached to the plains killifish's gills. Its name is *Salsuginus thalkeni*, the *thalkeni* being an honorific based on the last name of a landowner, Charles Thalken, who not only gave scientists permission to use his property for access to the river but also gave them a key to the gate.[11] This worm species was a newly discovered and thus newly described one, so the honorific *thalkeni* was a no-brainer. That's what biologists do in appreciation for gifts such as permission and a key, gifts that open both metaphorical and literal gates.

The body of *S. thalkeni* contains both male and female reproductive systems; the worm is hermaphroditic. The worms mate, with reciprocal copulation, and produce eggs that are released into the river and hatch into tiny ciliated larvae, which, if they are lucky, get sucked into the mouth of a plains killifish and encounter a gill filament, to which they attach with large hooks on their posterior end. This luck is evidently not influenced a whole lot by how much water comes out of the Colorado mountains. The average number of worms on a killifish gill and the number of fish infected remain about the same regardless of streamflow, month after month, year after year.[12] Statistically, one worm species is protected from those distant events that can decimate the populations of other inhabitants in their shared resource, the South Platte River.

Why do we believe this tiny worm and its fixed reproductive cycle avoid the boom and bust experienced by other parasites? The worm follows the fish, and it requires only the fish and a chance for the two partners in this relationship to encounter one another. When the fish retreats into calm backwaters during flood years, the worm, its eggs, and its transmission stages follow. The worm obeys nature not because it wants to or has been told to but because its evolved life history traits allow it, in fact force it, to do so. In addition, because its developmental cycle is direct, instead of indirect as in the case of *P. minimum*, it requires no additional resource, such as a snail intermediate host, that is extremely vulnerable to the river's fluctuations. The worm may be restricted to its immediate environment—the fish—but that environment tends to seek its own characteristic environment, to which the worm is adapted by default. The lesson is obvious: Behave in accordance with the way your essential resources behave, and you avoid the boom and bust. Your only real problem is to avoid being envious of the boomers while forgetting their busts.

If *S. thalkeni* could talk, delivering a political speech, its words would sound like liberal anathema to the modern Republican party: living in a manner consistent with its natural environment, its life not being controlled by the unpredictable availability of resources whose supply regulates survival; having respect for wild things that enrich its existence in a multitude of ways, for example, swimming into calm backwaters when floods boil down out of the Rockies; and not overwhelming its habitat by excessive reproduction. Of course, *S. thalkeni* cannot talk or, insofar as we know, think, look ahead, or plan for anything other than what its genes direct. The worm does nothing on purpose, at least as we humans define that term. It's locked into its life history by evolution, and the pattern of that existence makes it easy for humans to use it as a symbol, a model, a fable, that seems to teach us a lesson about our relationship with Earth: the fish is your environment; you have no long-term choice but to live how it lets you live instead of believing you can alter it to suit your needs.

From having studied these parasites for so many years, my morning news now sounds more like a lesson in parasite ecology than a natural disaster, military debacle, or political upheaval from halfway around the world. And yes, those decades of research are typical of mind-altering experiences that happen to scientists, especially parasitologists asking both how and why certain relationships are sustained during times of change. That's also the same question any of us could ask about events involving millions of people whose traits are relatively fixed, dictated by circumstances of birth, and whose environment is being altered, sometimes rapidly and irreversibly, by forces over which they have no control.[13] So a parasitologist reads the newspaper and thinks: in terms of potential impact, those forces operating on human affairs are not unlike the ones operating in the Pacific Ocean to make snow fall, or not, in Colorado, and worms flourish or fail—distant, for now, but eventually coming close enough to matter.

Such analogous thinking leads naturally to questions about other host-parasite systems, some of which might make good subjects for doctoral dissertations, some of which might be useful as lab exercises for freshman biology, and some of which involve parasitic ideas that end up changing the behaviors of their human hosts, behaviors that are not always beneficial or conducive to expression of our highest ideals. As a species, we build infectious agents out of words and pictures that we send throughout our own population with the goal of stimulating actions—believe this, now do that. Our information highway becomes a South Platte River equivalent, and sometimes those infective ideas affect our vision, changing the way we view one another or our common environment, and not always to our own benefit.

I will also admit that it was only after doing this research myself, cutting into those fish, quantifying what I saw through the microscope lenses, preparing specimens, analyzing data, and trying to interpret those results for all those years that I realized how much the lessons learned applied to me, my neighbors, and my nation. I read online the events of the day, I read books and magazines, and routinely I see examples of distant events affecting local lives, not only recently but also historically. I learn about elected officials, for example the 30 U.S. senators and 109 congressional representatives, who deny that the Earth's climate is warming and that human activities contribute to that change and who have accepted over $60 million in donations, even as the evidence for global change becomes overwhelming.[14] In some ways, they are like those killifish whose eyes are filled with *P. minimum* and who've never studied themselves or their kind over historical time periods, but unlike the fish, they have plenty of opportunity to get educated.

That changing climate, a planetary phenomenon, seems as distant to some of us as the southern Pacific Ocean currents must seem to the plains killifish, and distant not only in space but also in time. Corporate profits are calculated quarterly; lives, including those of children born today, are assembled over decades. Stock market prices are reported daily, if not hourly; ocean levels rise over decades, only three or four inches over the past thirty years, but scientists predict that when a child born today reaches their mid- to late seventies, in the year 2100, sea level could easily be seven feet higher than it is today. That prediction is a testable hypothesis, similar to a hypothesis we could construct now concerning worms in fishes' eyes, having done the research that establishes a relationship between global events and infections. That hypothesis would be a prediction of what our parasitological data would look like if we did that same long-term research again in western Nebraska. But this time in our analysis, we'd have to factor in what we've learned about climate change in the previous quarter century. It's entirely possible that instead of considering just El Niño and La Niña, we'd have to factor in carbon and methane levels in the atmosphere.

Those who deny climate change are also denying that our human activities have much if any impact on the environment in which we live. These men, decision-making and influence-peddling men, are playing the role of those worms living inside a fish's eye: they are obstructing our vision of what the world is really like. Their words are like parasites, making some of us blind to what Mother Nature is telling us through our thermometers and our rain gauges. In the meantime, climate change is estimated to have an impact on human health from combinations of flooding, malaria, diarrhea, asthma, premature births, crop failures, extreme weather events, and heat, with major impacts on the most impoverished, likely resulting in equally major political upheavals. Erratic weather patterns predicted as a result of ocean warming are actually happening. Permafrost is thawing, releasing methane that contributes to warming far more than does carbon dioxide. Smart real estate agents advise against buying coastal property.

Nothing we've learned from *F. zebrinus* and *P. minimum* is particularly new. Throughout recorded history, distant events have had major impacts on individual lives, with the familiar results being taught in schools ranging from elementary to colleges and universities—Columbus discovers the Western Hemisphere, again, for example, after the Vikings but with far greater effect on the aboriginals' individual lives. Whatever else they may have accomplished, the bombs dropped on Pearl Harbor on December 7, 1941, produced 180 million booklets of ration stamps, one of which my mother used to buy sugar in 1942. Those kinds of distant events and their local impacts are so familiar to us that we tend to think of them as the way things were instead of the way things are now and very likely into the foreseeable future, given the power we humans now have over transportation and communication.

Although the general question addressed by our parasitological research in a Nebraska river is how fixed systems deal with change, what we learn by trying to answer a specific question is what to look for when trying to answer this general one. Thus, I go home in the evening after analyzing data, pick up a copy of Karen Armstrong's *The Battle for God: Fundamentalism in Judaism, Christianity and Islam*, and can't help but make the connection: fixed systems do not deal gracefully with changes forced upon them by events beyond their control.[15] That assertion is especially true when those fixed systems are generated and sustained by humans who have a vested interest, usually involving power over other humans, in keeping those systems intact.

It may seem like a stretch, but I honestly believe that this issue of how established systems deal with change is not only a major one driving disruption and conflict in today's world, but it's also the reason that as a species, we're unable to deal with that change any better than can the worms in a killifish's eye. Although historians remind us that national boundaries have been redrawn throughout the centuries, when those boundaries are violated by people fleeing poverty and violence, for example, existing governance systems cannot always handle that border breakdown effectively and humanely.[16] That inability and perhaps unwillingness to deal with such changes rationally generates additional political conflict that in turn makes a big problem worse. Climate change, however, is the truly monumental change that our fixed human systems, especially those of governance, seem unable to address. If you need an illustration of that claim, consider that throughout the land traversed by the South Platte River and its plains killifish, the term "global warming" immediately closes so many local minds to humanity's most important problem for children born today, children who must eventually live with the consequences of those closed minds.

CHAPTER SIX

ROCKY MOUNTAIN TOAD

Mysteries of Transmission

Though boys throw stones at frogs in sport, yet the frogs do not die in sport, but in earnest.

Bion of Borysthenes, c. 325–c. 250 BC[1]

Once upon a time, in a sandpit pond far out on the western prairies, a large bass ate a baby Rocky Mountain toad, a little one, maybe an inch long with its legs stretched out. This tiny toad had just emerged from its tadpole life in a nearby puddle and made the mistake of falling into the pond, probably chasing some beetle. Then a young man went fishing and caught the bass, and because he was a student interested in parasites, he cut open the fish and started digging around in its guts. Finding the dead toad still relatively intact inside the bass's intestine, he naturally cut up the toad and started digging around in its guts too, in the process finding a young tapeworm. That discovery led to the death of 657 additional toads, especially baby toads like the one that had been eaten by the bass. All these little toadlets were just emerging from their tadpole stages and venturing out into the sandy world of the

South Platte River near an unincorporated Nebraska community named Roscoe when they fell victim to this student's scissors. Why did they fall victim? Because one of them had a tapeworm, and this budding parasitologist wanted to know how that worm got inside that toad inside that bass.

The toad collector rationalized his actions in a way quite familiar to scientists: he needs to answer a question about tapeworms that live in toads because discovering factors that inhibit or promote transmission will help him understand how infectious agents in general move through populations. Trying to find an answer will also force him to learn the statistical techniques that are used to analyze epidemics, regardless of what disease is being spread. Even if he doesn't solve this transmission problem but makes a valiant attempt to do so, some people will believe he's a real scientist, although others will think he's morally degenerate for dissecting all those little toads just to answer a question about tapeworms, not understanding that this question is not just about tapeworms in toads. But he's doing the job of a real scientist in a nation heavily dependent on science and technology regardless of what others think about his quest. Even though the subject is tapeworms in toads, whatever he's learned by trying to determine how an infective agent gets into and then moves through a population, he'll remember for the rest of his life. That memory will serve him well when and if, for example, he ends up in medical school and a deadly virus shows up in his nation, starts moving across the country, disrupts a nation's economy, and spreads death in its wake.

The question of how that toad got into the bass is a simple one; any child who's ever threaded an earthworm on a hook can answer it. Like many species of fish, bass are voracious predators, able to both see and smell potential prey, and a floating toadlet moving its legs is something any self-respecting bass would snap up instantly. That's why there are rubber frog lures in every tackle shop. But how do tapeworms get into toads, especially toads that young and tiny? And will those tapeworms stay with their toads year after year, growing and spawning more worms? With parasites, these questions are both fundamental and persistent; they will not go away. They're the first ones we ask and the last ones we must answer before trying to control infectious agents be they tapeworms, deadly viruses, or deadly ideas emanating from infected minds, minds without the vaccination provided by education.

If my conversations in private are a true indication of scientists' thoughts, we tend to believe that folks who might be incensed at this young man's toad killing are scientific illiterates, some of them willfully so. Understanding of infectious agents in general, including tiny tapeworms in tiny toads, is at a premium when fear and ignorance of science extend into the realm of public policy while a pandemic rages across the land. If you asked this bass fisherman with his toad tapeworm why we need to appreciate how his curiosity drives exploration, he would likely tell you that scientific illiteracy is a public health hazard. And he would be correct.

He might also quote a microbiologist named Hans Zinsser who, in his classic book *Rats, Lice, and History*, reminds us that pandemic disease has always been an enemy that cannot be vanquished by guns or posturing politicians.[2] Any hope of winning a war against germs or even against deadly ideas begins with the same question our young scientist is asking: How did it get there? If that question gets answered, he's ready to start on the next one: How does it move through a population of hosts and thus survive? A deadly disease or even a deadly idea cannot be stopped without an answer to that second question.

Those two questions—How is it acquired? and How does it move between hosts and thus survive as a species?—are related but separate. You can't answer the second without answering the first, but the first is typically a biological or in some cases a medical or veterinary question, whereas the second can easily become a political one, depending on what's infecting some hosts. It's only when we know the answer to the first but refuse to act on our answer to the second that we see how scientific illiteracy can become a public health problem. And a third question—How do we stop it from spreading?—cannot be answered without answers to those first two. Our young scientist with his bass, toad, and tapeworm questions will eventually become scientifically literate through experience and at a level surpassing by many orders of magnitude that of most elected officials.

How exactly will he acquire all this valuable experience? He'll devise some experiments that reveal what little toads eat and how much of it they eat. You don't get a tapeworm unless you eat something with an infective larva inside, so that means he needs to go to the river and think like a toad or act like a toad or both. He knows that his behavior will produce a human being who understands infection and transmission, regardless of who's being infected and what's being transmitted. He also knows that he can't study tapeworms unless he cuts up some toads, maybe a whole lot of toads, and that he can't devise experiments to test how toads get infected unless he has a supply of tapeworm eggs, which means he needs to also cut up some large toads, ones that have lived long enough to grow their tapeworms into adults. He doesn't know how long it takes for a larval worm to become an adult, with all its reproductive structures developed enough to breed, but he expects that some of those larger toads, ones that might have survived a couple of years on the river, will have adult tapeworms containing eggs.

The worm involved is named *Distoichometra bufonis*; it's long, thin, and somewhat delicate compared to some other tapeworms, such as the hefty and muscular ones in bass like our fisherman caught. The *bufonis* part of this name comes from the fact that this worm is found in toads, and the host was named *Bufo lentiginosus* when *D. bufonis* was formally described, in 1921, by a scientist named Lloyd B. Dickey from the University of Illinois.[3] This toad, commonly known as the American toad, is now named *Anaxyrus terrestris*.[4] Biologists never stop studying their subjects, asking questions about who's related to whom and what kinds of characters can be used to distinguish species. That's why scientific names end up being changed. The same thing happens to parasites, but so far, Dickey's name, *Distoichometra bufonis*, has stuck.

To date, nobody has returned to western Nebraska and collected these tapeworms for the express purpose of analyzing their DNA. Dickey described *D. bufonis* from specimens collected in Georgia, but the same species has been reported from across the country, most frequently in the arid southwest—New Mexico, Arizona, and Southern California—and in other toad species, for example the Sonoran green toad, *Anaxyrus retiformis*; the western toad, *A. boreas*; the Great Plains toad, *A. cognatus*; the American green toad, *A. debilis*; a tree frog, *Hyla cadaverina*; and even the New Mexico spadefoot toad, *Spea multiplicata*.[5] Whatever happens in nature to produce these tapeworms in toads and frogs is obviously happening across a wide part of the North American continent.

From what various parasitologists have told us about this tapeworm and given the molecular technology available to scientists nowadays, we know that a remarkable doctoral dissertation is living in toads just waiting for the right person to come along, do that

DNA analysis, and tell us whether *D. bufonis* has evolved along with its various host species, becoming diversified in the process. But that's also a dissertation that can be written, published, and used to launch the career of some young scientist without anyone ever solving the life cycle of this worm, figuring out how it is transmitted, and thus finding how the infections spread through the annual explosion of young toadlets just emerging from their tadpole stages—a yearly model for an epidemic in which thousands of naïve little hosts enter a new environment and immediately begin getting infected.

Throughout the northern Great Plains, toad production is an annual phenomenon. On warm spring and early summer nights, the sounds of Rocky Mountain toads calling one another to mate are punctuated by distant thunder, a prelude to rains that will supply tadpole habitats. Uncountable numbers of toad eggs will end up in relatively calm and shallow pools or marshes as long strings embedded in a jelly-like matrix. Equally uncountable numbers of tadpoles will develop from these eggs, eventually metamorphosing into toadlets that emerge onto moist ground and start eating any moving prey small enough to fit into their mouths. And every year, a substantial fraction of these newly emerged toads acquires their tapeworms. But one year, because of that fishing expedition, a bass, and an infected toad, a young scientist is about to embark on his two-year journey of discovery. He has plenty of material to work with; in June, while walking the river, he purposefully avoids stepping on newly emerged toadlets that seem to cover the sand. A garter snake moves through the weeds; that snake will eat more baby toads than our young scientist will ever use in his experiments.

Our parasite hunter who discovered the tapeworm remembers

and is guided by the two basic questions in parasitology, now phrased in terms of toads: How do these tiny toads acquire their worms? and How are these worms distributed among toads along the river, especially those that are now a year old or older? He really wanted to know not only how toads got their worms but also whether they kept getting worms as they grew during the summer and whether they lost them during winter when toads burrow down into the mud, their bodily functions are altered, and they're not eating anything to feed themselves and their resident tapeworms. As is the case with many, if not most, parasitic infections in wild animals, the second question was easy to answer—all one had to do was collect toads a year old, especially those recently surfaced from aestivation in the spring, and count their worms—whereas the first question turned out to be almost, if not totally, impossible to answer, sort of like the one involving *Nematobothrium texomensis* in chapter 1.[6]

But like most questions in science, even the easily answered ones, any attempt to answer them, successful or not, spawns additional questions. For example, one might ask whether the tapeworms have evolved physiological traits that allow them to survive in a toad's gut for the months their host spends buried deeply in the sand, its metabolism slowed through the Great Plains winter. We know those worms survive their host's time spent in the winter without food because in the spring, older toads come out of the ground infected. Suddenly we're talking about evolution in addition to ecology, and in doing so, we come away convinced that not only do evolutionary questions have ecological aspects, but the reverse is also true.

This relationship and interaction between ecology and evolution occurs with all organisms, but in the case of parasites, there are transmission stages involved, each with their own set of physiological requirements and each within an intermediate host that also has its own physiological requirements. A tapeworm egg is one transmission stage, and an insect that eats it, allowing that egg to hatch and develop into an infective larva, is not only what we call an intermediate host but also an environment required for larval development, like a uterine environment is required for a mammalian embryo to develop. The insect is an intermediate host because another animal, such as a bird, must eat that insect before the tapeworm larva can mature to an adult, now mating and producing more eggs.

We can thus assume that any traits that evolve to enable a larva to survive in some potential intermediate host will be of ultimate survival value to the species, including the adult worm in toads. Also, any traits that allow an intermediate host to survive in the toads' environment, whether that intermediate host is infected or not, will also be of ultimate survival value to the adult worm. And finally, there is the worm egg in toad feces, so any worm trait that promotes survival in the abiotic environment—temperature, moisture, etc.—or enhances the length of time that egg will be viable is also of ultimate value to the adult. For all these ecological and evolutionary reasons, it's never a surprise when some parasitologist starts talking about the multidimensional nature of parasite life cycles. Nor should it be a surprise to any of us if such talk makes us look at our own lives in terms of the complex array of factors that affect our existence, some of which are in our immediate environment but others of which may be happening halfway around the world.

The Rocky Mountain toad is now named *Anaxyrus woodhousii woodhousii*; at the time all this toad-collecting took place, it was known as *Bufo woodhousii woodhousii.*[7] Such names and their changes remind me of a discussion I once had with a group of students, including the one who first found a tapeworm inside a toad inside a bass, about what it takes to qualify for the title "minimal biologist." We defined the term as someone with the least amount of knowledge needed to support a claim they were a real biologist, deserving of the title. After a couple of hours, we reached the conclusion that you needed to know at least a hundred scientific names, the life cycles of those species named, the history of their discovery and description, their nomenclatural history—when and why their names had been changed and who had changed them—overall classification, evolutionary history, and biogeography, that is, their spatial distribution on Earth and the factors determining that distribution.

Most of us around the table could demonstrate this knowledge for a dozen or so scientific names; our requirement for a hundred was a self-imposed reminder of our ignorance, as well as that of most humans, about what lives on Planet Earth. You don't know what you don't know. The changing toad names also remind us that scientists are truly trying to discover evolutionary relationships, which we keep track of with official names and nowadays DNA sequences. We also adhere very closely to our established rules for naming living organisms because those names are data retrieval tools for access to scientific literature, sometimes from a hundred or more years ago. Our "thrown out on the table" number of a hundred names, life cycles, evolutionary histories, and biogeography was a reminder that even as scientists, we were still not very knowledgeable about Earth's biota, even the species we claimed to be studying.

Distoichometra bufonis, the worm that our young scientist found in the Rocky Mountain toad, could never be on any minimal biologist's list of a hundred species because its life cycle is simply not known and in fact may be unknowable, thus sharing a property with our worm in chapter 1. What is knowable about this worm and not the one in buffalo fish in chapter 1 is the rate at which their hosts, baby toads, eat potential prey items that could contain infective baby worms and the rate at which baby toads acquire their worms. We know these facts about how the world of nature operates because of our toad collector's research. If asked, the young man who found that first baby toad inside a bass would be able to explain how science is characterized first and foremost by the relationship between what is knowable and what is required before a possible knowable is transformed into a valid observation—something we call a fact. He's about to convert that explanation into tangible reality—numbers that tell the world something but not everything about baby toads and the baby tapeworms inside them. That conversion requires a couple of years and at least a couple of thousand miles of travel, some of it on rainy summer highways at midnight, where the grown-up toads can be found not only looking for mates but also shedding feces that contain tapeworm eggs. That conversion also requires an experiment, which in turn requires uninfected baby toads, lots of them, and thus a search for somewhere transmission is not occurring. Eventually, just by exploring, our young scientist lucks out.

Driving along a highway in western Nebraska, our bass-catching, toad-collecting, worm-seeking young man slows, turns into a sandy road, stops at the railroad tracks, then proceeds along that road until he finds a place to park. He gets out of his car, retrieves a five-gallon bucket from the back seat, and starts walking down to the sandy beach of a lake that's twenty-two miles long and three miles wide. Spring and early summer storms have been perfect for his mission: enough rain to make puddles where toads lay eggs, puddles that last long enough for these eggs to hatch and the resulting tadpoles to mature, metamorphose, and hop out onto the sand. The young man looks down. Newly emerged toadlets are hopping around by the dozens. An hour later, he has a hundred of them in his bucket. Eventually, he'll come back and get another hundred and another hundred; this experiment needs to be repeated before he can convince some anonymous reviewers that his science has been done correctly and should be published for the world to read.

What is he going to do with these tiny toads? The answer is very straightforward: he's going to do an experiment to discover the rate at which tiny toads get infected, but because he can't rear uninfected toads in large numbers in the lab, he must find a supply of uninfected ones, which he hopes he's done by looking for them somewhere other than along the river. When he gets back to his lab, he will dissect a third of these toads to discover if they are infected. If they do not have tapeworms, he will take another third of them, put them in plastic containers, and keep them alive until the end of his experiment. Why does he keep this group alive for a while? Because if the toads he immediately dissected are not infected, he needs to know if his toad supply from the lake stays uninfected. If that is the case, he assumes that anything he does to try to infect the last third of his toads will be a valid experiment. If this second group stays uninfected and whatever he does to this third hundred ends up giving them worms, he's made progress toward becoming a scientist, in the process acquiring not only a publication but also, and more importantly, transferable skills that will serve him well until the day he dies.

That progress will ensure that he never forgets the role that Rocky Mountain toads played in his professional life.[8] He also recognizes, however, that the design of his experiment is almost as transferable as the skills he acquired just by doing it. In the years after his life with toads, he'll see people trying to infect others with ideas, asking themselves what kinds of information, events, or observations might serve the role of intermediate host, thus enabling transmission. Those same people—candidates for public office are obvious ones—will be asking whether their target group became infected, that is, acquired new interpretations of their world. Every political campaign operates like somebody trying to do experimental infections with wild-caught hosts; the main difference between a candidate and our toad-catching scientist is that after the work is over, an uninfected toad group validates the experiment, whereas an unconvinced electorate invalidates a candidate's infection skills.

This kind of comparison and the metaphor come naturally to a parasitologist, especially during those long drives out to a research site when your thoughts tend to wander and you ask yourself why you are trying to answer a question that so few others might think is an important one. But the transferable skills acquired by doing this project are obvious even while you're struggling with the research techniques, logistics, efforts to find funding, and requirements of a formal graduate program—all models for the challenges you'll encounter the rest of your life. It's the experiments, however, that are truly educational. Our young scientist is asking himself: How do I solve this problem? What is a truly rational as opposed to emotional approach to problem solving? What kind of numbers and observations do I need to truly find an answer? That kind of life lesson in rationality is the main reason he's out on the river messing around with toads, and he knows it. If at the end of his experiment, there are worms in this last group—the third hundred, his so-called experimentals who'll spend the next three weeks in an enclosure out on the river—he will have conducted a valid experiment, although in essence, he will have allowed nature to do the critical part: infecting toads. The three answered and answerable questions are pretty straightforward: (1) Are my baby toads uninfected? (2) If I keep them only in the lab, do they stay uninfected? (3) If I put some in a corral out on the river, will they get infected by eating whatever comes their way in the night and will also fit into their tiny mouths? If the answers are yes, yes, and yes, our young scientist has made a major step toward solving a problem about how nature operates.

He still has a truly major step to take, however, and that is to explain that last yes. To even begin such an explanation, he will have to assume the persona of a baby toad out on the river. That's what scientists do; they start to think like their subjects. They also lie on their stomachs next to a muddy puddle and try to collect thousands of critters small enough to fit into a baby toad's mouth. To see most of those critters, you'd need a magnifying glass. Then you need to dissect them to find anything that might look like a tapeworm larva, and because the life cycle is unknown, neither you nor anyone else in the world has ever seen one of that species. So you're looking at a long expedition through a microscopic world searching for something you know exists but can only guess what it might look like.

Naturally, during casual discussions among local parasitologists—talk not unlike that carried out by John Steinbeck and Ed Ricketts while out on the Sea of Cortez—the conversation always seems to drift toward this story of tapeworms in tiny toads. And of course, talk about tiny toads inevitably leads to talk about tadpoles. Maybe it's the tadpoles that get infected, eating some aquatic tapeworm larva hidden away in a microscopic crustacean sucked into a grazer's mouth by accident. That's a question anyone who's studied frogs and toads would ask. Thus, our toad collector dissects hundreds of tadpoles, poking through their folded intestines and carefully examining every suspicious-looking piece of detritus through his microscope. Nothing, but he's not surprised. Tadpoles are generally vegetarians, with their long guts adapted to the difficult task of digesting mainly algae. Once they grow legs, their lungs become functional, and they move out onto the sand, these tiny toads transform into carnivores, with a much-shortened gut that digests prey items limited only by the size of their mouths. The transition from tadpole to toadlet thus involves a major change in gut chemistry, a change that can inspire some lucky larval tapeworm to start its growth into an adult.

But those mouths are tiny because the newly emerged toads are tiny. Our young scientist has already convinced himself that the tadpoles are not infected, nor are they the source of tapeworms for the newly emerged toadlets. So he starts his experiment, carrying rolls of aluminum border fencing and a shovel to the South Platte River, along with his bucket of newly emerged toads from that distant site. Three hours later, he has his enclosure, and in goes the experimental group. He'll be back to check on them daily. One garter snake, slithering over his fence, can eat his entire experiment. One great blue heron, courageous enough to land in that enclosure, could do likewise, all before his efforts lead to discoveries that confirm he is indeed a legitimate scientist driven by curiosity about how infectious agents are sustained in populations of the susceptible and approaching a problem in a truly rational manner.

He wonders how much baby toads eat, how many prey items need to go into those little mouths before one of them carries a tapeworm larva. That's a natural question for anyone studying parasite transmission, and he answers it by starving some of his little toads for a day, putting them into the enclosure, dissecting them at various intervals, and counting the prey items. He wonders whether these prey insects—and they are most certainly insects—will be a flying species or perhaps a springtail that for some mysterious reason ends up in his enclosure. Eventually he'll lie down on the sand and act like a baby toad, collecting everything that moves of its own volition, although instead of eating it, he'll take it back to the lab and dissect it, digging for worms in a microscopic drop of somebody's microscopic guts but never finding anything that looks like the larva he imagines. And eventually he'll step into this enclosure, collect all his toads, take them back to the lab, and do the same thing he did with that one from the bass: dissect them, then dig through their guts for worms.

What did he discover by doing these experiments? Newly emerged toads eat about six prey items an hour; anything that comes within an inch of that little carnivore and is small enough to fit in its mouth is probably snapped up instantly. Over the course of weeks, he dissects members of the experimental group, in the process discovering small tapeworms. Just by putting uninfected toads in an environment where he's collected infected ones, these immigrants get infected. By the end of his experiment, his numbers show that these tiny toads acquire about one tapeworm a week. He knows that when he goes to the river in the middle of the night, it's easy to find toads of all sizes hopping around on the sand, just like it is during much of the day. He makes some calculations: 6 prey items an hour times 24 hours a day times 7 days a week equals 1,008 prey items. On average, a newly emerged toad less than an inch long, not counting its stretched-out legs, gets infected with one tapeworm for every thousand prey items it consumes.

This eating and acquiring occurs all summer long. Something that goes into that little mouth has a larval worm inside it, but nothing our young scientist has done or will do during the next year tells him what that something is. He attempts experimental infections using his abundant supply of tapeworm eggs, which he feeds or tries to feed to every insect he can culture or capture that is small enough to fit into a toadlet mouth. None become infected; none are candidates for the source of tapeworms in his tiny toads. His dilemma is the same as was John Teague Self's with his worm from the ovaries of buffalo fish along the Oklahoma-Texas border. The dynamics of seasonal infection can be described, but the basic cause of infection—the transmission mechanism—is a problem of such magnitude, played out at microscopic scale, that it resists discovery. What seems so logical as it happens on the South Platte River sand cannot be brought into the laboratory.

But when our young scientist leaves the river and goes out into the world, he carries that lesson with him and recognizes it in places far removed from his research site, including his daily news. We humans can describe our problems easily; solving them is another matter altogether, especially when there are others involved. That is a major life lesson from a parasite. But it's the serious attempt at actually solving a problem instead of just complaining about it or blaming others for it that produces the lesson learned. And because we've learned that lesson from experience, it sticks with us and shapes our reactions to those who complain and blame instead of recognizing that yes, we as a species do face some daunting problems just screaming for the same kind of rationality forced on our young scientist by tadpoles, toads, and tapeworms.

That relationship between a problem's description and its solution permeates human affairs, never more obviously than when a nation is invaded by deadly disease, people fleeing violence, or the effects of a warming global climate, and the weapons to handle these invasions are built from knowledge and the rationality that produces it instead of from steel and explosives. The first step in using knowledge and rationality, however, involves admission that our beliefs and desires can blind us to reality. And furthermore, it's not just one small problem with parasites that seems so contrary but problems splashed across our media outlets day after day. Current challenges facing my nation from across the Rio Grande, for example, include not only people seeking safety and opportunity but also invasive plants and disease vectors following a warming climate. But it's the human migration that seems to end up in the nightly news. If there is a rational discussion about immigration, you're more likely to hear it in your living room than in Congress; similarly, the wealth of available data on the fate and role of immigrants is far more likely to show up on your computer screen, driven there by you eager for a history lesson, than in a local newspaper.[9]

In the final analysis, if this adventure with bass, toads, and tapeworms has accomplished anything for the nation in which it occurred, it is the production of one person who can easily admit that nature is behaving counter to his desires, that the solution to this mystery of how toads get worms will eventually require more knowledge, rationality, and effort than he has expended so far, and that now he sees this relationship between a problem and its solution in places far beyond a river or a lab. Ideally, this lesson could somehow be passed along to those holding public office or hoping to, where it would end up being applied to problems of far more significance to our nation than tapeworm transmission mechanisms. Maybe someone needs to convince this young person to run for public office.

CHAPTER SEVEN

DEATH OF A BEETLE

Who We Kill, Why We Kill, and Why It Matters

Kills on contact!

Kills up to 12 weeks

Directions for use

From a spray can purchased at the store

When biologists deal with regulations surrounding the use of animals in teaching and research labs, especially when those animals must eventually be killed, their bar talk often refers to the way political rhetoric can reduce the emotional impact of killing our fellow humans, rhetoric that we hear on certain news stations and read in our daily newspaper. For Americans in the 2020s, the words "Russian," "Taliban," and "drug cartel" function in that manner. Institutional concern over the death of a mouse, a regular topic in biologist conversations, seems never to be matched by those responsible for delivering lethal weapons to hot spots of human conflict around the world. Parasitologists seem to be inordinately occupied with animal care and use regulations, primarily because we usually need to dissect vertebrate animals to recover their worms or assess the pathological impact of an infection.

Biologists' talk about animal death also routinely includes comments about the millions of birds killed annually by feral cats or that die in collisions with city buildings. So I'm not surprised when walking across campus one morning on the way to class, after I'd just written a lab exercise in which students would need to kill mealworms to discover how infective agents exist in a population, lo and behold, there is a dead warbler at the base of a building, having flown into a plate glass window and broken its neck. I stop and pick it up. It has not been dead for long; rigor mortis has not set in and there are no ants around the eyes. I take it to class, intending to pass it around among my 260 students in BIOS 101. Of course, such use of a dead bird in class is against all the rules of my institution and probably also federal law, but you never pass up a chance like this one—never—just because of rules and laws.

What should one do with a bird in the hand? The answer depends on what kind of bird it is; if injured, how seriously; whether it's dead or alive; if dead, how long it's been dead; and whether there are other people nearby. My general rule is to put any bird in the hand up against your cheek, although I resist if it's a species that will peck your eye out. But you should do something that will forever implant the feel of that bird into your brain; holding it against your cheek works best. Whenever we humans use more than one of our five senses—seeing, hearing, smelling, tasting, and touching—in an encounter, the multiple use tends to embed the memory of that encounter more strongly in our minds than if we used only one of the five. The feel of that body and those feathers against your cheek will change forever the way you think about birds. If the bird is alive and makes a sound, that memory is instantly branded on your brain; but even if the bird is dead, that close encounter will stick with you for a very long time.

If the encounter is with a small bird, like the dead warbler now in my hand, alive but fatally injured, most parasitologists will kill it by squeezing its chest cavity, thus suffocating it and stopping its heart, as shown to me by a famous ornithologist. Later, those parasitologists will look for worms in the intestine—evidence for an infection carried on a migration. Fortunately for me, this bird is already dead, so I won't have to demonstrate that technique, although had it been alive with a compound fracture of the wing, common sense about the politics of higher education and the regulatory environment I work in would have led me to kill it before bringing it to an audience of first-year college students. Carrying the warbler, I enter a building, joining my students in the hallway. Some of them notice the bird. Inside double-glass doors, I walk down the auditorium aisle to my pontification place, start flipping switches and tapping touch screens to activate the electronics, and put this warbler on the document camera.

The warbler is now an image five or six feet long up on the screen. When I describe what happens to birds when they hit windows and how a professional scientist deals with a mortally injured but still living bird like the one I'm holding, there is laughter from the audience. I was not expecting laughter at my description of injury and death; that was a true shock. For most of these young people, a personal encounter with death awaits them in the years to come—that of grandparents or parents or in a few cases that of a classmate, by accident, suicide, or military conflict. There will be no laughter at these times. Some of those in my audience are avid hunters and will, when deer season arrives, take their high-powered rifles and sit in a blind with the intent of killing a large animal. Will they laugh when that bullet enters a deer's chest? Probably not; for them, this is serious business. Hunters are supposed to kill deer.

A tiny bird felled by a campus window, however, is a part of nature rarely experienced so closely, and this morning, I have an opportunity to teach something far more valuable than whatever my PowerPoint show provides. I hold the warbler against my cheek, demonstrating the technique for ensuring one never forgets such an intimate encounter with a wild thing, a ten-second lesson in beauty and vulnerability. Then I offer this opportunity to the students seated in the first row. Nobody accepts it; some of them seem to shudder at the idea. My audience is the future of my nation and my species. Their reaction to death and the beauty of this bird will not leave my mind, even as I deliver some standard lecture on genetics for the next forty-five minutes. Some of these students had to be from small towns; they could not all have been from suburban Omaha and therefore so separated from every aspect of nature that they'd shudder at the opportunity to hold a dead warbler and remember the reason it died.

Back in my office after class, I sit down and read the lab exercise I have written, in which students must kill beetle larvae known as mealworms to learn about parasites, and compose the following essay. Of all the pages that I have written about death over the past decades, this particular essay, the result of a dead warbler at the base of a campus building and my decision to use it as a teaching device, seemed to survive the test of time. So here it is, "Essay on the Death of a Beetle":

> In his truly magnificent bestseller book, *Lives of a Cell*, Lewis Thomas, nationally acclaimed cancer researcher and president [at the time] of Sloan-Kettering Cancer Center in New York, included a chapter titled "Death in the Open." He begins this chapter with a discussion of road kills described as fragments

of familiar animals.[1] His essay addresses death as a natural phenomenon and ends with comments on humanity, reminding us that within fifty years of his writing, our population will have doubled and with this increase we'll certainly see the reality of human death, about fifty million a year.

Whenever I develop an undergraduate laboratory exercise that involves death of an animal, even a beetle or an earthworm, and especially one in which students are assigned the task of doing the killing, Thomas's words come back to me, along with those of E. O. Wilson (*On Human Nature*) and Paul Fussell (*Wartime*).[2] In his chapter on aggression, Wilson talks about the dehumanization of fellow humans as a prelude to violence, especially in times of social conflict. Fussell is more explicit, using WWII as an example, and citing ways in which we dehumanized enemies, thus desensitizing not only our soldiers but also the folks back home. In my Field Parasitology course at Cedar Point Biological Station, in which we routinely sacrifice animals to discover "who's infected with whom," the basic observations necessary to analyze any parasitic relationship, I often end the semester with an extended discussion of Thomas, Wilson, and Fussell, as well as some more modern cases involving massive human destruction (Rwanda, Kosovo, Persian Gulf War, etc.).

There is a simple reason why I often feel that such a discussion is necessary: When you come to know an insect, snail, or "minnow" rather intimately and build your reputation on the scientific study of their parasites, then it is not so easy to dehumanize these lowly creatures. These organisms, with which you do your first real research project, show someone you are truly capable of conducting an original scientific investigation, and earn your guaranteed get-into-med-school letter of recommendation, suddenly become valuable to you. They are no longer worthless trash, they are no longer repulsive, they are no longer something you have absolutely no feelings whatsoever for, but instead they become a part of your emotional and intellectual library. They've given their lives, yes, but they've also given you analytical powers, the irreplaceable power of experience, and the intellectual sophistication that comes from doing research, that you would never have been given had you not set about to study their parasites.

We chose beetles this week because we grow them in large numbers, they are not endangered, no permits are required for their use, and they are not like us [furry, warm, with large dark eyes]. For most of you, this week's lab will be the first, and perhaps the only, original experience you will have with the distribution of infectious agents in a population until you graduate from medical school, get into practice, and deal with a flu or head louse epidemic. If this prediction turns out to be true, then I hope you remember your lessons well. And if, as a "health care professional," you find yourself caught up in a military adventure, then you will probably also find yourself wishing you had studied the biology of infectious organisms over and over again and been somewhat less enamored of reproductive physiology, cancer, and cardiovascular function.

This discussion leads, of course, to my rather smart-aleck comments about dead birds at the base of city buildings and my perhaps unwise advice to simply pick up a stunned bird with a compound fracture of the wing and kill it. Those comments were intended to accomplish one thing and one thing only: to vastly increase your sensitivity to death at the population levels and put into some kind of rational perspective our use of beetles this week in lab. Actually, I was a little bit shocked at the class's reaction to those comments; I did not expect laughter when I described how a world-renowned scientist that I knew killed birds by squeezing their lung cavities. By way of comparison to the migratory bird situation, about 45,000 Americans die each year of gunshot wounds.[3] Another 42,000 die in automobile accidents.[4] From a biologist's perspective, especially a biologist who studies small organisms, the clearing of tropical forests at the rate of 50–100 acres *a minute* for the past 20 years results in the death of uncountable but truly beautiful and wondrous organisms.[5] Fourteen thousand deer were struck by automobiles in Iowa last year, at a cost of about $3000 per incident ($42 million a year in damage).[6] A friend of mine who regularly rode a bicycle along a country highway and counted road kills, then extrapolated that sample to the national level, estimated that at any moment there would be 75 million birds lying dead on America's highways. I read a report (unconfirmed) that house cats in Great Britain killed an estimated 60 million songbirds a year. The Kearney arch has cost one human life, and not too many years ago the *Omaha World-Herald* reported that the increase in speed limits from 55 mph to 65 mph on I-80 resulted in approximately one additional human life a month. The speed limit is now 75 mph.[7] A visit to a packing plant makes your hamburger and bacon look quite different from before the visit. And, of course, I have not addressed the issue of quality of life for those still living who, for various reasons, do not have access to the humanizing influences of quality education, a safe place to sleep at night, adequate health care, and meaningful employment. Into this latter category fall millions of Americans and billions of other human beings around the world.

I'm not condemning anyone for contributing to the above figures; I am, however, simply reminding us that just by living our normal, twenty-first-century, human lives, we contribute to the death of vast numbers of organisms and generally ignore the deaths of vast numbers of human beings, all except, that is, the ones closest to us. Thus, it does not bother me very much to use beetles to provide young people, many of whom will become physicians, with their first scientific experience with infectious organisms [we all have nonscientific experiences with infectious organisms].

On a more personal level, I do appreciate the fact that an intimate encounter with death, as when you cut the head off a mealworm or separate a beetle's head from its body, can produce an emotional reaction. In this particular case, you have chosen to terminate a life in order to study something that most people find repulsive (a parasite), even though that repulsive organism is living the most common way of life on Earth. I only ask that you remember this week's lab when your kid comes home from day care with head lice or pinworms and you wonder how to cure the infection; it's not very difficult, at least in the case of pinworms.

Finally, as a philosophical aside, as part of your overall education as a biological sciences major, I strongly recommend a personal examination of your own reasons for reacting as you do to the welfare of other organisms, be they insects or fellow humans. I'm guessing that the closer an organism is to you personally, or the closer in appearance and demeanor to humans in general, or the younger the organism, then the stronger will be your reaction to its death. This principle figures prominently in politics and government regulation surrounding the use of animals in research and teaching. Thus, the death of a baby cocker spaniel has an infinitely higher emotional content than the death of a mosquito or cockroach, at least to the average person. And if you contribute to that death, then the puppy will probably linger in your mind for a lifetime, whereas the mosquito and cockroach will be forgotten as soon as you get over the pleasure, and probably smug satisfaction, of having killed them.

My lab exercise, intended to demonstrate how infective agents are distributed in a population, involved killing and dissecting mealworms. We can buy mealworms online and at local pet stores, either alive, frozen, or freeze-dried, or mixed in suet blocks for backyard birds. These larval beetles are thus commercial products, reared to be eaten, although not necessarily by people, but nevertheless in this regard they are like the billions of chickens that humans consume each year. By eating agricultural products such as corn, we contribute to the death, by poison, of uncountable numbers of insects. So just by living in the United States, especially if we are anywhere near the imaginary "middle class," we contribute to a massive death of organisms with which we share the only planet known to support life anywhere in the universe. With that perspective, I hope you're thinking it's okay for college students to tear off a mealworm's head and start digging around in its guts for parasites just so they can have an original and relatively safe experience with infection before a viral pandemic arrives in their neighborhood. Hopefully, the experience of generating original data will also be of later use when they encounter "news" that ignores relevant numbers involving some political issue or delivers rhetoric that stirs emotions without solving any problems.

But that act, carried out under a microscope on a live and squirming victim, does not have the emotional protection provided by packing plants, grocery stores, and restaurants. Nor does it have the endorsement of a hunting and fishing tradition promoted by a local tourist industry. Never in those students' experience have they been asked to kill something with their own two hands just to make an original observation, get a grade, and learn a lesson. So when I ask them to dissect a living insect, it's always with the hope they'll discover parasites and that such discovery will generate enough curiosity about infectious agents in general to help them rationalize the emotional impact of what they've just done to discover some. I also hope that the ensuing discussion of infection will help them understand the movements of other agents—including viruses and strange ideas that alter a host's behavior—when those agents arrive in their own communities and infect people they know or may have voted for in a local election.

How do you kill a mealworm for parasites? The easiest way is to simply use forceps as cutting tools, grabbing the larva right behind its head with one pair and using a second pair to slice the head off. Actually, it's more like ripping the head off. When that happens, of course, there is a lot of squirming, so you grab the mealworm's rear end with forceps, holding the decapitated front end with the other pair so it doesn't flop all over the place, and gently squeeze the rear end. If you do this decapitation and squeezing correctly, the gut protrudes out the front end, whereupon you again use your forceps as a cutting tool, ripping off the rear end, so that the whole gut, now unattached, can be pulled out intact into a drop of water on a slide. Then you need to gently tease the gut apart, releasing a swarm of one-celled animal parasites called "gregarines"—a term derived from "gregarious"—referring euphemistically to their mating habits.[8]

What do you see in this drop of water after teasing apart the gut, still using your sharpened forceps as cutting tools, adding a cover glass, and putting the slide under a microscope? My answer, from having done this kind of work for decades, is that you see one of the most captivating scenes in all of biology. You see growth, development, commitment, loneliness, diversity, and sex. You also see, depending on how many species are in this mealworm, comparative morality: some kinds of these one-celled animals are waiting until they are old enough, signaled by being large enough, before mating, but with other species, mating can occur between the very youngest (well, maybe not the *very* youngest, but close) as well as between the young and old. And having observed all this mating, with the anterior end of one cell embedded in the posterior end of another, the question arises of which is the male and which the female or whether we should just call them "plus" and "minus." Hard-core cell biologists might ask questions about proteins on the surfaces of these cells, but no prof at the podium of a class with 260 young people in it would pass up the chance to talk about sex, even between microscopic cells, an opportunity delivered by a dead beetle larva.

All it takes is a scale inside a microscope lens to convert this lab exercise into numbers that can be analyzed statistically by counting and measuring who participates in single-celled sex inside a beetle and who doesn't. Mealworms have three species of these parasites, distinguished by size and shape, all living in the same gut but at somewhat different locations and exhibiting different mating habits. We could easily add another species of beetle to the exercise, further extending this analysis into consideration of how environment shapes mating habits, although in this case, those so-called habits are evolved instead of learned. So now, because we've killed beetles and extracted their guts, we can ask about the relationship between where parasites live and their mating habits and whether those habits are evolutionary adaptations to something we humans can never discover or understand. Class discussions are likely to be interesting.

It's a real temptation at this point to let your mind wander into the culture wars that seem to sweep across our nation unabated by public education and probably exacerbated by private education. It requires no imagination to understand that the abortion wars are basically a heated political debate about sex instead of about a collection of cells' potential life as a toddler. If I were a lab instructor supervising this exercise involving parasites of beetles, it would be a real temptation to ask that students assign parasite species to political parties and explain their decisions. If we had enough time, it might even be fun to ask whether parasite mating habits could be altered by changing the beetles' intestinal environment, for example with different kinds of food. Would oatmeal instead of wheat germ "teach" that flagrant species to behave "properly" and choose mating partners in a way we humans might consider relatively constrained? I'm guessing that it would be very easy to write a lab exercise using mealworms and their parasites and asking for interpretations of data that would produce a complaint to the Board of Regents and a faculty disciplinary letter once that lab was discussed during Thanksgiving break at home.

If you're that lab instructor, on the other hand, you're probably lucky that these parasites are not arguing over something the mealworm should be doing to protect them from parasitologists, for example, crawling to the bottom of the culture instead of lying around on top where a person with sharp forceps can pick them up easily before tearing off their heads. And you're doubly lucky that these parasites cannot talk to those students, given that the conversation might involve destruction of homes, intrusion into privacy, withdrawal of life support, immersion into alcohol or other toxic chemicals, manipulation, exploitation, and ultimately death, all delivered as a result of some human's curiosity, a student trying to answer a question about parasites, and doing it in a way that will teach them how to deal with more serious infectious agents a decade hence.

If you're a scientist looking through this microscope, however, instead of all this sex and morality issues among parasites, you see publications, theses, and dissertations, all swarming around in an insect's guts. The proverbial person off the street sees single cells; a parasitologist looks through a lens and sees a career characterized by continuous study of infectious disease on a global basis, the multitude of vectors involved, socioeconomic factors that influence health, and the impact of parasites on agriculture. Developmental events, transmission mechanisms, and evolutionary relationships—all these seemingly academic subjects are tied closely to the leading question of our time, namely who's infected with whom? The "who" can be any of us; the "whom" can be anything from head lice and pinworms to goofy, even wacko, conspiracy theories about everything from viruses to election results. Any biologist would put a warning label on toy microscopes: Keep your kids away from these things if you want them to be Wall Street traders.

Some kid's story of wanting to become a scientist because of a birthday-gift microscope is so close to reality, so close to what I have seen happen time and time again as a result of dissecting mealworms, that in my mind, it's more of a prediction than a story, more a way of teaching life lessons instead of just a lab exercise. I've seen that career focus happen because of the questions a college prof can ask about parasites in insects and the opportunities for research that same prof can provide in a laboratory. I'll admit to looking at all my students as potential scientists, pursuing an honors thesis because of what they saw inside a beetle larva. And there was a large supply of those young humans, the vast majority of them not really knowing what kinds of experiences they might have in the years to come but expecting that people like me would deliver some meaningful ones because they were paying plenty to be in my company. Or, as it turned out, in the company of mealworms.

The act of teaching about movements of parasites through populations is itself an attempt to infect a population with knowledge that will change behavior. And not only do we parasitologists consider teaching a form of infectious agent transmission, but we also figure out ways to make those agents of understanding more virulent. Those of us who stand in front of young audiences, sometimes with a dead warbler in hand, sometimes with mealworms on our mind, hope that at least some of the people in those audiences will understand what is happening and try to apply that understanding to resolve the next major public health problem. That hope is not confined to animal parasites; it extends to parasitic language that can alter the conditions under which millions of our fellow humans exist. If that hope is realized, a beetle will not have died in vain, and that warbler will never be forgotten.

CHAPTER EIGHT

AN ANTI-DESENSITIZATION PLAN

Lessons from Death

MUST BE REGISTERED TO HUNT BUY OR SELL SNAKES
12PM DEN OF DEATH AND BUTCHER SHOP OPENS

www.okeenesnakehunt.com[1]

A few years ago, a scientist who was studying parasites that lived inside reptiles, birds, and mice did something that thousands of nonscientists do routinely: kill reptiles, birds, and mice. The problem was that this scientist and the students who worked under his supervision evidently did not kill these reptiles, birds, and mice in an approved manner. This killing, especially of a rattlesnake, was supposed to be done according to rules established by a committee appointed to oversee how people collected, handled, cared for, injected, measured, weighed, experimented with, and killed (euthanized) animals and disposed of the dead bodies afterward.[2] To understand the rest of this story, it's important to know that none of the people on this committee had ever killed an animal just for the worms inside, to understand how infectious agents moved around the world, and possibly not for any scientific purpose at all.

Before exploring the result of this unapproved killing of a rattlesnake, it's probably also important to know that American scientists use at least twelve million animals a year in research, the vast majority of them rats, mice, birds, and fish. To quote the website speakingofresearch.com: "We use fewer animals in research than the number of ducks eaten per year in this country."[3] I don't know if that number of ducks eaten includes those shot by hunters or only reared for food, but if that website information is correct, U.S. residents eat more than twelve million ducks a year.

Mice are the most commonly used vertebrate animals in research, but genetic manipulation of mice, including selective breeding, has greatly reduced the use of other species, such as dogs and cats. But it's also important to understand that the larger the animal, the more it costs to breed and maintain it. Scientists are no different from the rest of us in trying to reduce costs, and if they can use a genetically modified mouse instead of a beagle for some experiment, that is a massive savings. That savings includes the emotional impact of killing a dog or cat, or even a nonhuman primate at the end of an experiment, if necessary for scientific reasons.

How do scientists use animals in research? The answer is in very many ways. Every physiological measurement that the average human is subjected to during a lifetime of health care can also be performed on mice or rats without killing them. Every blood chemistry analysis, every cardiac response to exercise, every test of vision and smell, every diet's impact on weight gain or loss can be done for research purposes on mice or rats without killing them. The vast bulk of animal research is aimed at improving human welfare and that of our pets and livestock. Infectious disease drives much of this work; the nearly five thousand research projects initiated in Germany after March 2020 and focused on SARS-CoV-2 (the virus that causes COVID-19) used nearly eight million animals.[4] We don't know how many German mice died of COVID-19 as a result of human efforts to develop a vaccine that probably would have indirectly saved the lives of some anti-vax crusaders.

Some kind of animal use typically qualifies a person to be on such a committee; behaviorists and physiologists are often chosen, although sometimes philosophers or local ministers are added, presumably to ensure that scientists understand ethical issues. But no one on this particular committee had ever dug around in a dish of guts just to find a tapeworm that might be a species new to science. None had ever asked how parasitic worms were maintained in populations of mice or, for that matter, in buffalo fish along the Oklahoma-Texas border as described in chapter 1 or in toads along the South Platte River as described in chapter 6. When the supposed departures from approved killing methods were reported, anonymously of course, by what we might call a whistleblower, all hell broke loose and the scientist's career was drastically altered because of how he went about looking for worms.

This scientist was a teacher, and his subject was parasitology. His classes were among those of extreme value to college students destined to become physicians—some if not most of whom would be dealing with infectious agents, trauma, death, and even pandemics for the rest of their careers. The committee revoked the scientist's permission to continue teaching his parasitology course in the previous manner. The judgment was so vindictive that this scientist was not allowed to even pick up rabbit or field mouse feces from the prairies so he could teach students how to search for cysts or worm eggs—the simple basic techniques used globally to assess the impact of parasitic infections on human populations. Among the demands for reinstatement of permission was the development of an anti-desensitization plan. There were no guidelines given as to what such a plan should contain.

When I learned about this requirement, I contacted the disciplined scientist and we had some lengthy conversations about the draft plan he was preparing and the reasons for including certain references and readings. Until this committee action, I had never heard of such a requirement; it had never occurred to me that such a plan would ever exist. A Google search on the term "anti-desensitization" turns up about six and a half million hits, with virtually all of them devoted to *desensitizing* a patient who's developed a hypersensitivity, usually to some prescription drug.[5] In the psychology literature, the term also describes a treatment that reduces an emotional response to something. Thus, the scientist was being asked to develop a plan to prevent his students from becoming insensitive to the collection, killing, and dissection of wild vertebrate animals, even though some of those students were avid hunters and fishermen.

Before COVID-19 managed to shut down the nation's colleges and universities in the spring of 2020, I came to the campus where this anti-desensitization plan incident took place nearly every day. I would walk over to the student union, get some coffee, and retire to a site where I'd spend an hour of creative writing and eating a couple of pieces of dark chocolate. I'd been doing this little routine for about forty years. One day as I was walking across campus, I encountered the person who was chair of the disciplined scientist's department. He'd not yet heard of the committee's action. During our conversation, we talked about how this revocation of permission to collect would affect this prof's ability to teach about infectious agents and what the result would be relative to premed majors who would eventually encounter a pandemic. I mentioned the anti-desensitization plan requirement. The chair was rather taken aback.

"Is there a model for such a plan?" he asked. Even as an experienced scientist with a long career, he'd evidently never heard of such a document.

"A national model?" I responded. "I don't believe so."

It was only a few minutes later, with my coffee, chocolate, and open laptop, that the concept of an anti-desensitization plan began to metastasize through my thoughts. So many people in my nation had become desensitized to the plights of their fellow humans that we needed a plan. Naturally, that term—"anti-desensitization plan"—took on a life of its own and became material for this chapter, especially so because my own work as a scientist had also involved permission from this same committee in question to collect and kill small animals, including fish, lizards, frogs, toads, and mice, and cut them up to search for parasites. However, no permission is required to kill other kinds of animals, including beetles, snails, and mosquitoes. If you were somehow able to personally identify with a mosquito, considering it close enough to humans or beautiful enough to elicit an emotional reaction from humans other than entomologists so that it needed protection by a procedure and ethics committee, we'd consider you mentally ill instead of qualified to be on that committee. The same claim could be made about the killing of roaches used in research.

The last time I filed a permission request—termed a *protocol*—with this committee, I needed to complete a seventeen-page form describing exactly how I was going to do these acts and how the animals were to be handled from the time they were collected until I disposed of their remains. Also, at one point in my fifty-year career, I was asked to fill out a medical form, upon which it seemed like every other question was about whether I had an allergy to cats; evidently such a reaction might be an institutional liability, thus to be avoided. But until this conversation with my fellow scientist about his denied protocol occurred, I had never heard of an anti-desensitization plan. Neither had he. We concluded that it was a made-up idea, something that might have popped up in a local bar, but instead it was generated in a committee meeting and an agenda item involving regulatory compliance.

Naturally, because we were both professional biologists and parasitologists, my conversations with this disciplined colleague turned to the number of vertebrate animals killed annually by various activities in which we humans, including those on animal care and use committees, engage in regularly—for example eating, driving, working in buildings with glass windows, and having cats as pets. And just as naturally, our conversations involved published scholarly estimates of those numbers. So here are some figures to consider the next time you go to the grocery store, take a driving trip on the highway, enter a tall building, or let your cat outside for some fresh air. All these estimates are well supported in the scientific literature. The numbers tell us about our relationship to animal life on Earth, whether that life be in a tropical forest, a prairie, or a factory that produces chickens. We humans kill so many animals for so many reasons that the ones sacrificed to learn how parasites move through populations are hardly noticeable, except, I suppose, to a committee never exposed to a tapeworm's allure.

Data from 2013 reveal that domestic cats, both pets and feral ones, kill somewhere between 6.9 and 20.7 billion mammals and 2.4

billion birds every year.[6] That's "billion" with a "b." And that's every year. The birds are mostly species protected by federal law because most of them are migratory ones. A large fraction of all these mammals and birds are infected with parasites, including tapeworms, roundworms, single-celled organisms, fleas, lice, and ticks. Every year, feral house cats destroy more parasites, along with their host wildlife, than a whole nation of parasitologists could identify in a dozen lifetimes, parasites that if they could be studied seriously would be fodder for thousands of scientific reports, most of them providing information to help us understand how infectious agents are sustained in nature.

It is estimated that motorists kill almost 400 million animals annually; yes, annually. That toll includes an estimated 41 million squirrels, 26 million cats, 19 million opossums, 15 million raccoons, 6 million dogs, and 350,000 deer.[7] Every year. Some highway killings are intentional, especially when the victims are snakes and the drivers are male. And as is the case with feral house cats, more parasites are destroyed along with their host wildlife than would be required for thousands of dissertations on the biology of infectious organisms, the generation of which would provide the world with a massive supply of scientists who truly understood disease transmission and could teach elected officials, at least those with any interest in learning, about the risks to our economic systems from infectious organisms before the next deadly one arrives on an airplane.

Figures from 2016 to 2019 tell us that we slaughter about 50 billion chickens a year (not counting male chicks and unproductive hens which, naturally, just get killed and their bodies disposed of somehow), 1.5 billion pigs, a similar number of goats and cows, and half a billion sheep.[8] Brazil has the world's largest population of cattle and to sustain it has cleared a staggering amount of Amazonian forest, destroying in the process untold millions of birds and their nesting habitats, reptiles, small mammals, and uncountable numbers of insects, rendering some of them extinct.[9] The scientific expertise that could be produced by focusing on the biology of parasites in Amazonian forest fauna, were it not destroyed by clearing for agriculture, cannot be estimated. But that expertise could easily help us understand how exotic diseases show up in human populations. While the Brazilians are producing lots of cattle, by erasing tropical forest that could teach us plenty about ecology, evolution, and sustainable agriculture, they are also producing ignorance about the only planet known to support life anywhere in the universe because we're losing any opportunity to study the organisms that live in those incredibly diverse forests, assuming scientists could gain access to them.

Humans consume 150 metric tons of seafood every year, about half of it in the form of cultured fish and shrimp.[10] Collection of ocean fauna, including fish species not used for food but killed in the process as well as invertebrates, matches tropical forest destruction in terms of the number of scientific studies that could be done on these host species and their parasites. Just by living on Earth, the present population of between seven and eight billion humans is obliterating not only the planet's biota but also opportunities for the intellectual endeavors and understanding that this biota could provide if it were readily available to scientists, including young ones looking at careers teaching others about how our planet's ecosystems operate and what makes these ecosystems thrive or collapse.

On a much smaller scale, whatever else this animal use committee may have accomplished with its anti-desensitization plan directive, it guaranteed that many university students would be ignorant about a lot of the parasites they might encounter in the future, especially those found in mammals. Most of the students taking this professor's class will end up as physicians and other healthcare professionals having never gained the appreciation and knowledge of how a tapeworm lives and moves within our world. That knowledge could easily be of real value when used to help patients years hence, and not only with possible tapeworm infections. A broad knowledge of animal parasites is an acquired trait that can be applied to all infections, regardless of the organisms causing them.

The numbers mentioned above do not include turkeys, ducks, geese, guinea pigs, and various other species kept for food, like the pet sloth I was allowed to play with while on a trip through the Peruvian Amazon a few years ago. My best description of that sloth, its feel, and the way it clung to my arm is "darling"—sensuous, dependent, like a baby. When I asked the young girl who had the sloth what she fed it, she replied, "leaves"; when I asked what she planned to do with it when it grew up, she said, "eat it." That answer put her in the same category as people throughout the world who eat "bush meat" or buy recently captured wild mammals in an open market and slaughter them for food or use in traditional medicine practices, risking infection by parasites and viruses that were perfectly at home in the wild but could easily spread through the human population.

Pangolins, for example, were originally notorious for the belief or maybe just the idea that they were the source of SARS-CoV-2, the virus that managed to shut down much of the world's economy in 2020 and, due to evolution of new strains, continued to send thousands of people to the hospital every week for the next year. Pangolins are captured, destined not only for food but also scales (actually, keratin, like hair and toenails) that supposedly have medicinal properties, at the rate of about one every five minutes.[11] To my knowledge, none of them are examined for parasites. Every five minutes, a pangolin is dissected for food and scales instead of by someone satisfying their curiosity by digging in its guts for worms. As famous as pangolins might be as models for infectious beliefs, they were eventually denied their role as a source of SARS-CoV-2, an obvious case of lost fame. Pangolins, however, are not the only scaly animals collected by humans for various reasons, including food.

In northwest Oklahoma, there is an annual event centered at the small community of Okeene (population 1,149) called "the Annual Okeene Rattlesnake Hunt," or, as I remember it from growing up in Oklahoma, the Rattlesnake Roundup.[12] This event includes several days of entertainment and collecting; meals of snake meat, of course; and Boy Scout pancake feeds. In April 2019, a member of Cowboys from Hell Snake Handlers broke a record by lying in a bathtub with 211 rattlesnakes. This celebration of collecting and killing is billed as the oldest organized snake hunt in the world, with eighty-one years as of April 2020. I don't have an estimate of how many other snake hunts are held annually, where they are held, or how many snakes they harvest, but the Okeene roundup typically results in the killing of at least a thousand pounds of rattlesnakes. I've eaten rattlesnake, captured in south Texas while I was on a field trip to collect migratory birds as part of the Bird/Virus/Parasite project (chapter 2); the snake was okay fried in some beer batter, but the idea of eating a rattlesnake was better than the meat itself.

Given the prior discussion of Okeene's annual rattlesnake hunt, this is probably a good time to report that one of the main reasons our parasitologist scientist was denied permission to even pick up a rabbit pellet to search for worm eggs was that he put a rattlesnake in the freezer to slow it down before killing it according to protocol (injection of a narcotic) but got busy and forgot the snake until it had frozen to death. Writers, of course, have imaginations; mine is busy building a picture of this scientist at Okeene, Oklahoma, in April, telling the Cowboys from Hell Snake Handlers what happened to his career and why it has happened. And after that, he's telling them that because of the way he killed this rattlesnake, he now must write an anti-desensitization plan and get it approved by a committee of college profs before he can kill another one by some method approved by this committee. My guess is that they'd direct him to the Snak Shak for some fried snake meat.

Regardless of how unwelcome they might be to the snakes involved, for ecologists and herpetologists, organized rattlesnake hunts can be a bonanza of scientific information. For example, in the late 1980s, with the participation of University of Kansas researchers, the Oklahoma roundup produced some excellent data on rattlesnake diet. George Pisani and Barbara Stephenson analyzed the gut contents of 347 freshly collected adult snakes and reported a rather nonselective predation on local rodents, including ground squirrels, gophers, and nonnative Norwegian rats, as well as moles, cottontails, and a jackrabbit.[13] Snakes don't chew their food, so it takes a pretty big snake to swallow a jackrabbit whole; the Norway rats suggest either that the twelve snakes eating them had ventured into a local garbage dump or that these typically domestic rats had made it out into the Oklahoma hinterlands where they encountered the predators.

Did these snakes have parasites? Yes, of course, including a species, *Porocephalus crotali*, belonging to the Pentastomida, or tongue worms, now understood to be crustaceans with a fossil history going back to at least the late Cambrian, five hundred million years ago.[14] Although he collected his rattlesnakes in the Wichita Mountains of Oklahoma instead of around Okeene, John Teague Self, the scientist so consumed with the life cycle of *Nematobothrium texomensis* as told in chapter 1, also found pentastomes, a largely ignored group of parasites at the time. Self's subsequent fascination with this group eventually had the same effect on his mind as did those worms in buffalo fish ovaries, but the results were far more successful, at least scientifically, as he ended up one of the world's experts on tongue worms, traveling the world to collect and study them, often collaborating with other scientists.[15] One of my letters from him, dated July 7, 1969, reads:

> *Pentastome work warms up all the time... The day after Labor Day I take off for Africa and ultimately will visit Nigeria, Kenya, Bombay, Bangkok, Kuala Lumpur, Taiwan, and Okinawa returning home the 28th of September.*

Research on the parasites of snakes earned Self enough of a scientific reputation to make him welcome in labs around the world; I seriously doubt whether he ever had to file a protocol with some committee to collect these snakes or collaborate with those scientists in Africa, India, Malaysia, Taiwan, and Japan. Teague Self was a dignified, highly ethical, and sensitive individual, even when the business was collecting wild animals, but knowing him as I did, I hate to think of the articulate sarcasm that would be produced if some committee demanded that he write an anti-desensitization plan, then rejected his best efforts repeatedly. He would have been irritated but patient with the initial demand; the repeated rejections would have inspired some words not fit for mixed company. As an aside, in Texas, it's routine to spray gasoline into snake dens prior to collection; collateral damage includes the impact on other reptiles such as collared lizards, a truly dramatic member of the southern Great Plains fauna, which may lose its ability to forage effectively for a week. I'm certain, without checking, that any amateur snake collectors in Texas asked whether gasoline was an approved chemical compound to use on snakes would return a look that cannot be described in writing, probably accompanied by equally indescribable language.

The plan my colleague had to prepare, with no guidelines, could easily have been the first ever by a scientist and certainly the first ever by a scientist who intended to study parasites. For the record, the plan was rejected, as were subsequent editions and revisions all made in response to succeeding demands by this committee. The college science course involved is titled Field Parasitology, and it is taught at the University of Nebraska Cedar Point Biological Station. So here is the essence of this unique document prepared by a scientist. As you probably suspect, there was no way for me to read this plan without deciding to eventually write a similar anti-desensitization plan for my nation. You should also know that when this scientist included text extending the concept of anti-desensitization to humans and providing literature to support this inclusion, the committee made him remove that text and those supporting references before rejecting it again.

Here we have an example of a committee more concerned that students might become desensitized to the killing of mice than to be reminded that they could also remain or at least become desensitized to the killing of their fellow humans that happens routinely around the world for various reasons, including politics, religion, pathologically insecure men, and access to natural resources. The nearly fifty thousand gun deaths in 2021, for example, over half of them suicides, seem not to have sensitized very many elected officials enough to support controls, although the forty-three thousand traffic fatalities that same year suggest driver's licenses and insurance requirements don't necessarily inspire people to handle dangerous equipment safely.[16] And war, of course, no matter who's participating or when and where it happens, involves a desire to kill others, as dutifully reported by the media.

How effective is a anti-desensitization plan? Would such a plan prevent you from becoming desensitized? Or would it make you more aware of what you were really doing and why you were doing it, assessing the cost to nature against what you intended to do with the lifetime's worth of education you're receiving about the movement of infectious agents in a population? To answer those questions, I suggest imagining yourself as a student starting to work on a thesis involving infectious worm transmission in a population of toads (chapter 6) and being required to have on file an approved protocol before cutting open that first tadpole, as was the case during those years. But given the fact that with your training you also have the potential to become a university scientist studying parasites, thus eventually being asked to assure a committee that your students are not desensitized, it's probably important to consider what might go into this as-yet draft document. If I were writing one for my own research, and a committee had removed all references to the desensitization that accompanies war and political actions on immigration, I would probably file a formal complaint with some other committee, one that handles cases involving academic freedom. You can't discuss becoming desensitized to the death of a rattlesnake or a mouse without also bringing up what happens when nations decide to destroy one another, especially if that decision is made by some megalomaniac "leader."

Nowadays, largely because of the speed with which information travels, it's impossible to ignore violence that happens throughout the world. For the same reason, it's also impossible to ignore public commentary about this violence—words and ideas coming from news sources, public officials, and citizens using their First Amendment rights. Historians remind us that humans have been notoriously destructive throughout the past few millennia, although that reminder tends to be smoothed over by analysis of the results—nations born, kings ascending to power or being replaced, and geographic boundaries redrawn. But my own use of animals in research and submission of protocols for approval greatly heightened my awareness of death on its many scales, from that dismembered insect to shootings and whatever war is currently being waged somewhere. Yes, I honestly believe that use of animals in research has made me significantly more aware of violence, death, and destruction than I might have been otherwise. I also believe that every scientist who uses animals quickly develops that same awareness, along with a deep appreciation of what their research reveals about life on this only planet known to be occupied by sentient beings. If I were teaching a course in parasitology today, I'd use my colleague's anti-desensitization plan, approved or not by some committee.

The plan developed by our real scientist included general statements about the reasons for using vertebrate animals, the rules that governed their work, thorough training in the handling of those animals once collected, procedures for preserving tissues and recording data, trapping methods, the rationale for using a variety of species, and training in the proper preservation of parasite specimens. Expressly forbidden was the keeping of any parts of any specimens for personal reasons; nobody walked away from this field experience with a rabbit's foot for good luck, even though such useless tokens can be bought from a variety of local places and online sources.

Also expressly forbidden by the disciplined scientist was the naming of animals collected for study; naming is what we do with pets. These animals already had names, scientific names, and once in the lab, they also acquired a collection number. Scientific names designate membership in an identifiable group, catalog numbers mark individuals, and official records establish a rationale for the research. Scientific names and specimen numbers indicate that because this specimen goes into a museum collection, it is of value well beyond any immediate educational use. This anti-desensitization plan also included readings: historical literature about the desensitization of humans to the killing of other humans, people belonging to an identifiable group and thus given a number, but in this case a number tattooed on their arms. The committee removed these literature references from the draft plans submitted. During conversations with the disciplined scientist, neither of us could come up with a reason for the removal of these references and readings from the associated literature. Both of us had referred to such literature in classes for many years.

What was this literature that the committee felt was not apropos? The list of reading materials included not only scientific papers supporting statements about numbers of wild animals killed by domestic cats, for example, but also more general literature about life on Earth, the way we study it, the reasons we study it, and the lessons we learn from letting curiosity drive our efforts to understand natural systems. I was honored that our disciplined scientist chose the foreword from one of my own books, *Dunwoody Pond: Reflections on the High Plains Wetlands and the Cultivation of Naturalists*, as recommended reading.[17] That book is about the origin of young scientists as exemplified by people whom I watched grow intellectually from college students in Biology 101 into practicing, publishing scientists and college profs teaching classes full of premeds. I promise that these students' exposure to parasites of vertebrate animals was a major lesson that heightened their awareness of all kinds of potential health issues.

I truly believe that their use of vertebrate animals, in addition to insects and snails, was a key factor in this heightened awareness. From my own personal experience as a parasite hunter, experience that includes watching hundreds of college students pursue research projects involving animals ranging from microscopic invertebrates to insects, frogs, toads, birds, and mice, I honestly believe the plan developed by this scientist was an excellent one that addressed the potential problem of desensitization, if it is a problem. Of particular importance was his statement on the naming of animals: *don't do it.* If you name some living organisms, for example your pets, they take on a special character, a companion-like property that is surprisingly close to the way you view other humans.

You can see where this discussion is headed: to understand part of nature, you must collect it, dissect it, and preserve parts that also preserve that understanding—tangible evidence for what the world is really like as opposed to what some elected official is telling you it is like. You must also compare your original observations and discoveries with those reported in what's called the primary literature, scholarly journals in which publication requires anonymous reviews by experts. Unlike the names you give to your pets, scientific names and museum specimen numbers provide a means of data retrieval and, if justified, the borrowing of such specimens. For scientists, those official names and numbers establish respect for the subjects of our research because of what these organisms teach us about life on Planet Earth.

Yes, it is possible to trap an animal somewhere in the world, cut it open, and spend the next six months preparing and trying to identify the parasites that host contained. That is a familiar experience for parasitologists. It's the "trying to identify" part of this activity that can be educational beyond description, a powerful reminder not only of what's involved in documenting Earth's resources but also how unfamiliar is so much of what lives on the planet with us. That identification exercise also produces valuable life lessons because of the artistry required for specimen preparation, microscopy required to decipher internal structure, the struggle of finding and dealing with appropriate literature, the experience of responding to anonymous reviewers, and the final satisfaction of seeing your work in print, especially if you've discovered a new species. Everything you do from the time you set traps to the day you read that publication with your name on it sticks with you for the rest of your life and gets applied to problems far removed from parasitology, for example assessing what kind of evidence supports outrageous claims of political candidates. That's a life lesson experience the committee denied so many of this disciplined scientist's students.

The aforementioned teaching and research activity is quite different, indeed quite the opposite, from collecting parts of nature for profit, as in the clear-cutting of old growth forests in the Pacific Northwest or strip-mining in Appalachia. Maybe if we named each of those eight-hundred-year-old Oregon spruce trees or each lump of coal from that West Virginia mountain, such naming would be the first step toward development of an anti-desensitization plan for our environment.

CHAPTER NINE

PROFITING FROM PARASITES?

Why It Could and Maybe Should Happen

Banknotes, being objects of great turnover and diffusion among the population, may be efficient mechanisms in the dissemination of various intestinal parasites.

Marina Costa, Layane Teodoro, Gustavo Bahia-de-Oliveira, Ana Nunes, and Ricardo Barata, *Research and Reports in Tropical Medicine*, 2018[1]

Parasitologists often try to figure out ways of managing and killing parasites, with varying degrees of success. For example, they thought the biochemical action of avermectin, a roundworm-killing molecule produced by bacteria discovered on a Japanese golf course, could be improved, an idea that led to the development of ivermectin, with subsequent release of the patent by Merck and Company. This kind of thinking and humanitarian action enabled the treatment of millions infected with *Onchocerca volvulus*, the parasite causing severe skin disfiguration and ultimately river blindness when juvenile worms released into the bloodstream eventually reach the eyes and die, producing inflammation. The importance of this drug and Merck's actions produced a Nobel Prize shared by the developers—Satoshi Ōmura and William Campbell.[2]

Ivermectin is also used in deworming dogs and horses to prevent heartworm infection and intestinal parasite buildup that can lead to illness and even death. It should go without saying that ivermectin is not a cure for COVID, which is a viral illness, but evidently the data on this issue does not convince some people, including a few physicians.[3] Still, we learned that soil bacteria can give us wonder drugs that work against some infections, at least in the case of ivermectin. We also learned that an unusual idea can infect people as easily, maybe even more easily given the electronic vectors involved, as a real virus.

Attempts to develop a vaccine against malarial parasites have been less triumphant than efforts to prevent river blindness, but not because scientists haven't been trying. Only after decades of research by some of our finest minds, with demonstrable success along the way, has a vaccine been developed that seems to have potential for reducing the impact of this parasitic disease. But malaria remains a major health and economic burden throughout much of the tropical and subtropical world, especially in sub-Saharan Africa where children are at greatest risk. Great ideas can encounter equally great resistance from Mother Nature, but such resistance rarely stops scientists from trying to pursue those ideas, especially if the payoff is profound, as it was in the case of ivermectin.

Thus scientists, particularly those with interests in the control of infectious disease, never stop imagining what we might accomplish if able to bring seemingly weird ideas to fruition. For instance, parasitologists sometimes wonder how we might make something that functions like the seemingly indestructible transmission stages, technically called oocysts, produced by parasites in insects such as mealworms. These oocysts are less than 4/1000 of an inch long and when viewed with an electron microscope look something like this:

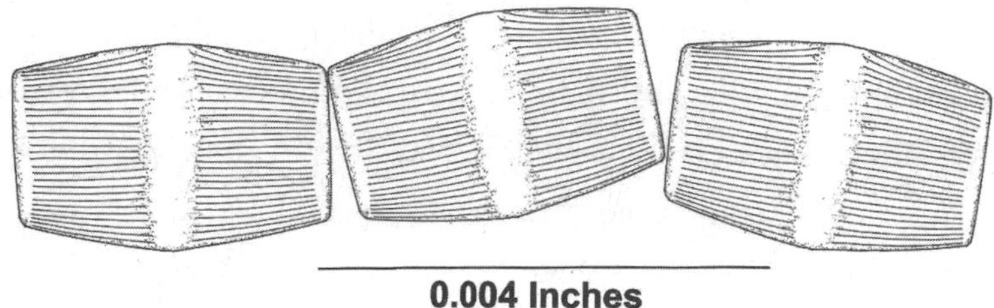

Each of these microscopic containers carries eight infective cells that emerge when the cysts are swallowed by mealworms, thus infecting or even reinfecting a host. Such cysts have been produced by the uncountable quadrillions for millennia and likely consumed or breathed in by every person who ever stored and ate grain where insects thrived, playing a hypothetical role of the perfect drug or vaccine delivery vehicle if we could only make them perform according to our wishes instead of some parasite's needs.

What might those wishes be? The answer is anything that reduces the cost of maintaining and delivering drugs and vaccines, especially to the far corners of the world. For example, COVID vaccines are relatively fragile, being shipped from the manufacturer at somewhere between−58°F and 5°F, thus requiring major refrigeration equipment, including that involved in transportation, until the vials arrive at any of the thousands of local pharmacies for injection into someone's arm.[4] Those vaccines are good for another month in the equivalent of a household refrigerator, assuming the temperature is at the recommended 40°F or less, but for only a day if left out on the shelf. People lining up at their local pharmacy for a "free" COVID shot likely never ask the person delivering it what's involved in making sure the fluid in that vial has not expired, but if they did, the answer would be a serious lesson in transportation economics, with the price of diesel fuel being a minor factor.

In 2019, a group of authors from the nonprofit ThinkWell Global and the Bill and Melinda Gates Foundation published a paper titled "The Costs of Delivering Vaccines in Low- and Middle-Income Countries: Findings from a Systematic Review," in *Vaccine: X*, a peer-reviewed open-access journal.[5] Vaccines analyzed ranged from PCV-10 and PCV-13 (pneumonia) to HPV (human papilloma virus), measles, yellow fever, hepatitis B, DTP (diphtheria-tetanus-pertussis), and others. The authors' data search, retrieval, and quality assessment were remarkable for their thoroughness and strict criteria. Thirty-three countries were included in the analysis. As a minimum, costs included health worker training and salaries, transportation, refrigeration, equipment depreciation, building utilities, supplies such as syringes, printing, and waste disposal. And the final figure? Excluding the cost of vaccines, the final figure for one fully vaccinated child ranged from $8.36 to $96.16, depending on the country, with an average of $40.90. The countries included in this study contain 4.8 billion people, well over half the global population.

Now comes some "what-if" thinking that approaches science fiction. But remember that there are plenty of people alive today who witnessed the origin of hydrogen bombs, moon landings, laptops, smartphones, oral polio vaccines, organ transplants, the internet, Google, Twitter, and COVID-19 vaccines, just to mention a few of the technological wonders delivered by the global scientific community since the end of World War II. Wouldn't it be nice if we could build not only a vaccine for malaria, containing, for example, cells that could build antibodies but also a package that would not require refrigeration, could be mailed anywhere for the cost of a postage stamp, and would stay viable on the shelf, without refrigeration, for a couple of years?[6] It turns out that such a package exists, but it's only made and used by parasites in insects instead of by human chemical engineers and healthcare providers.

Production of those cysts is only one example of a very long list of functions that nonhuman species, including parasites, perform but humans cannot. For example, we cannot digest wood like the protists in termite guts can do. We can't jump a hundred times our body length like fleas or carry twenty times our own body weight like ants. Nor can we survive for two years without food or water, like the cysts produced by parasites of beetles that thrive in places where humans store grain. Besides all that, we cannot even store grain without attracting a variety of insects (like mealworms), most of which are infected with parasites at some stage in their life cycles. If we could discover how these parasites perform certain tasks, like producing cysts that keep their infective cells alive for years—again, without refrigeration—that discovery could easily help humanity deal with its own infective agents. We ship all kinds of materials around the world, including drugs and vaccines; it would be nice if we could do that as easily and cheaply as mealworm parasites "ship" their infective stages.

The beetles whose parasites produce these cysts have likely been thriving wherever grain was stored throughout human history, a good example being Bronze Age silos dug into rock on the island of Malta.[7] Even today, somewhere in the world, beetle larvae are eating grain intended for humans and thus also feeding the parasites within those larvae. Current estimates of stored grain loss due to insect infestations range from 5 to 10 percent in the developed world up to 40 to 50 percent in less developed regions.[8] Among those insects, a relatively large and thus obvious one is *Tenebrio molitor*, a species of darkling beetle we now call yellow mealworms, a name based on the color and behavior of the larvae. If you need an introduction to the uses humans might have for a pest, visit the nearest pet store and pick up some suet blocks filled with these larvae, grown commercially. Woodpeckers love those kinds of blocks. Or, if you have a pet lizard, buy it some live mealworms.

We don't know whether the prehistoric Maltese considered mealworms in their grain to be pests or if they ate them on purpose; archaeological evidence from fossilized feces in other parts of the world suggests the latter possibility. But any species that grows easily in readily available food, like mealworms in cereal, are a gift to both teachers and researchers nowadays, another good example being the flagellated protists from scuttle flies, as described in chapter 4. In the millennia since the Maltese dug their grain storage silos, the yellow mealworm has worked both for and against the human species, although its role in human affairs has evolved from an agricultural pest into food for exotic pets and backyard birds, partly because we've developed methods for reducing its numbers in stored grain, at least in some parts of the world. The yellow mealworm has also been the subject of graduate student dissertations, as well as ideas about using parasites for profit.

In the realm of animal culture, if scientists are trying to supply themselves with organisms to use in research, this word "pest" has certain positive connotations, although it's not certain that same judgment can be made of bad ideas that function like pests, growing, often malignantly, in places where they are not particularly welcome. It's this combination of growth and place that leads to our use of the term "pest," but when a species seems to multiply easily without a lot of tender loving care, is small enough to keep in the lab in large numbers, and does something biologists find interesting, you can almost be assured that species will end up as the subject of someone's research. The yellow mealworm, one of the insect species that's probably invaded stored grain since humans started storing seeds, is a perfect example of this principle.

Also, unlike hosts for pathological ideas, mealworms come with no regulatory burden to inhibit your imagination, no committee making sure you cut off their heads correctly, no forms to fill out before you disembowel them to get their parasites, and no annual report on how many you killed while trying to make a significant discovery of some kind. Even young scientists quickly learn that regulatory burdens will be a normal part of their work, and whatever they can do to reduce those burdens will help maintain the fun and excitement that accompany research. Get these folks together in a bar and they'll complain about anything that seems to detract from their work—a scholarly habit that's probably spread across all disciplines. But when the conversations concern parasites in stored grain pests like mealworms, the complaints are more likely to be about choice of lab music than about regulatory compliance.

Whenever I listen to parasitologists talk, I also think of John Steinbeck's famous description of invertebrate zoologists at work and play in and around the Sea of Cortez, equating biologists to tenors because of their behavior and demeanor.[9] Steinbeck may have been thinking of himself and his friends with this description, but when comparing biologists to tenors, he also reminds us that tenors have the greatest range of male voices. Breadth is a critically important asset not only when stepping into certain opera roles but also when dealing with complex problems of any kind, especially parasitism and the kinds of ideas that can end up being discussed when parasitologists are at play, for example at their favorite Friday afternoon watering hole.

In college town bars, conversations can easily turn to potential uses for the microscopic cysts produced by parasites of mealworms, parasites that probably infected mealworms since the evolutionary origin of these beetles back in the Mesozoic and produced uncountable quintillions of such cysts in the hundred million years since. Such cysts were obviously consumed accidentally by not only those ancient Egyptians eating grain from their underground silos but also every other animal that ate seeds or licked off its paws after chasing a mouse into the granary. Those mice, the cats chasing them, and the Egyptians too probably breathed in those cysts. But to profit from these cysts, we need to figure out how to control their fate and function, purposefully using something a parasite makes to transmit something that humans make, for our purposes vaccines against a virus, and distribute it intercontinentally with no refrigeration required.

The cysts produced by parasites of mealworms are tiny packets of infective cells. Technically, they are known as oocysts because their protective coverings are formed around a fertilized gamete, that is, an ovum. The parasites belong to a very large group known as gregarines, which parasitize vast numbers of insect species as well as earthworms, marine annelids, and a variety of other host groups. Gregarines are transmitted by these oocysts, which, when eaten by the proper host species, release their infective cells, which then grow to maturity, usually in the host's intestine. Achieving maturity, the parasites mate by fusing, then secreting another cyst wall, becoming a gametocyst, and within this gametocyst, "female" and "male" gametes form and then fuse to produce fertilized ova. These fertilized ova secrete their own cyst, the oocyst, within which they multiply to produce more infective cells and ultimately are spewed into the environment. If lucky, the oocysts are consumed by just the right host and the cycle continues.

Given what we now know about these parasites and their cysts, a typical parasitologist's question about ancient Egyptians might be: Did they also get infected with the parasites that live naturally in mealworms, and if not, why not? Parasite cysts, like parasite eggs or even parasitic ideas that someone has picked up somewhere, get distributed throughout any environment in which an infected host resides or travels. Exchange of parasites between wild animals and humans is common enough that we have a word for it—zoonosis, in the singular, and zoonoses, in the plural—and the species that harbor such agents also have an appropriate name—reservoir. There is plenty of evidence, for example, that the COVID-19 pandemic is a zoonosis;[10] other zoonotic agents have names familiar to us from the travel literature: Zika, West Nile, Dengue, Ebola, and, of course, from history—plague, caused by *Yersinia pestis*, an infection we can share with various rodent species.

Closer to home, people who keep cats as pets are strongly advised to empty the litter box daily, and not just because of any smell. Cats are notorious reservoirs of *Toxoplasma gondii*, the cyst-producing causative agent of toxoplasmosis, a disease that can cause fatal brain infections in newborn infants by way of movement across placental membranes.[11] House cats, like mealworms in the lab, were taken from nature sometime in the distant past and now live with people. The main difference between cats and mealworms is that the cat's parasites are infectious for humans, whereas parasites from a mealworm's gut are not. In fact, the parasites of mealworms are rather cute, resembling the Shmoos that showed up in *Li'l Abner* comic strips many decades ago.[12] And not only are they cute, they can also perform almost like Shmoos did, although they provide humans with all kinds of research opportunities, leading to our understanding of parasite transmission in general instead of offering themselves up for meals that taste like chicken, beef, pork, or catfish. They don't lay eggs like Shmoos, but mealworm parasites do produce cysts, their transmission mechanism, and those cysts in turn inspire ideas that hatch in scientists' minds, ideas about the purposeful transport of various items important to people.

The mealworm parasites' inability to infect us is a result of their chemical makeup, a product of their evolutionary history. I am reasonably confident of this conclusion because humans study humans, extensively, medically, behaviorally, socially, and culturally, all in addition to whatever other perspectives a pathologically narcissistic species can bring to bear on the question of who they are, how they exist, and why they exist. We know that at least 260 different kinds of eukaryotic parasites—those having membrane-bound nuclei within their cells—have been reported as infecting humans.[13] These organisms range from amoebas that live in people's mouths to twenty-foot-long tapeworms writhing around in their small intestines to head lice crawling over their scalps, and their effects range from hardly noticeable, at least in healthy individuals, to lethal. Never once, since the origin of scientific literature, has there been a report of these Shmoo-like parasites from mealworms occurring in a human.

Since that first person ate a mealworm, humans who've eaten grain have likely been passing these parasite cysts in their feces, sometimes because they ate the beetles or larvae but most often because these cysts were in their cereal grain food, deposited by the mealworms and surviving whatever food preparation methods have been used in the past millennia. We suspect that either humans digest the cysts or the parasites do not survive what we call excystment—analogous to hatching—because throughout history, human feces have been examined in far more detail than the average person can envision. Most of that examination involved a search for parasites, and never, since microscopes were invented, has a scientist or physician observed one of these beetle parasite's cysts in human feces. Or, if observed, its presence was not consistently associated with disease. So a parasitologist might focus on the question of what would happen if you could control the fate of a mealworm parasite's cyst, maybe even inside a human being. What property would we focus on to make our millions in the pharmaceutical industry? That's a good question, and we have an answer.

To date, you can't control the ultimate fate of that cyst except in two ways: feed it to another mealworm or kill it. These parasite cysts can be rendered noninfective with heat or time, the former requiring a hot plate or oven, the latter requiring years. The project these parasitologists are about to discuss involves the years and another option: take advantage of one amazing property of these cysts, unravel the biochemical secrets that provide and sustain this property, adapt that property to deliver pharmaceuticals to humans and domestic animals, patent the adaptation, and watch the money roll in. The scenario you've just read has been followed thousands of times by scientists, and some of those attempts have been successful. The last time you filled a prescription, you were the beneficiary of these scientists' accomplishments. But you've yet to be prescribed a pill with its ingredients protected by the same covering that protects a mealworm's parasite. Ideally, that pill could contain a miracle drug to erase a bunch of humanity's problems, ranging from cancer and old age to any of the other many ailments you see mentioned in televised ads for pharmaceuticals.

Cysts from mealworm parasites are truly tiny, smaller than a human's red blood cell; have a structure that makes them relatively un-wettable; and can be blown like dust for as far as a wind can carry dust particles. The infective cells inside are called sporozoites.[14] The other thing we know about these cysts is that they remain viable for two years on the shelf. We know this fact because of experiments conducted by a student named Richard Clopton, who eventually acquired a PhD and became the world's expert on these kinds of parasites.[15] And of all the discoveries made by Dr. Clopton, the two-year viability of cysts sitting on a lab shelf leads to what might be the wildest idea of all: the commercial and medical use of these cysts if certain discoveries could be made about them. In other words, can we convert a parasite from a pest into a medical device of global importance? If that sounds like an outlandish question, remember that a molecule produced by bacteria from a Japanese golf course ended up as ivermectin, delivering a Nobel Prize to a couple of scientists who developed it, then gave it to a parasitized world.

For example, when you collect these cysts from mealworm feces, then feed them to the right kind of insect two years later, parasites will emerge and continue their lives inside the insect's gut—a microscopic version of the Rip van Winkle story, except these two years would be about seventy parasite generations instead of a short twenty years for humans. In this case, the parasites are so harmless that people who study harmful species tend to snicker when you mention that you are doing research on gregarine parasites in mealworms. Ideally, to gain the respect of your self-important colleagues, you'd probably need to demonstrate that these parasites kill insects. Then they could be used for biological control; but that's not the case. There's no evidence that these parasites do anything more than use some of the mealworm's food and grow into cells that look like miniature snowmen. Eventually these little snowmen pair up, secrete a cyst around themselves, and reproduce like crazy, making thousands more cysts that then exit the mealworm's gut with feces. And there they can lie for up to two years, waiting to be eaten or blown into a field half a continent away, preserving the infective eukaryotic cells within them in viable condition.

It's the "infective," "preserving," and "viable" parts that make the "two years" an important dream, an incipient technology just waiting, indeed begging, for development, then application. In other words, the parasite is doing something humans cannot do, and these parasites are doing it every day, by the quadrillions, moving on wind and water as far as that wind and water will take them and just waiting for the right insect to eat them but hedging that wait by surviving for two years in that package, the cyst, holding eight viable eukaryotic cells. And that package is made from proteins built according to parasite genetic instructions. That fact alone should make molecular biologists' imagination run wild the next time they walk into the lab, stare at their DNA sequencing machine, start wondering what problem to tackle next, and start imagining what would happen if they took Gene A from Species X and put it into Species Y.

Suppose, for example, you could replicate that cyst wall, along with the factors that control its ability to recognize a specific environment, then release its contents. Suppose you could do everything this parasite does but do it with eukaryotic cells of your own choice, genetically altered to perform some function, instead of those microscopic parasite cells. In other words, do what scientists regularly do nowadays, but package the results in a cyst that behaves like the ones from mealworms. What you need to extract those millions of bucks from the human population is the ability to do something that so far only parasites in insect larvae can do, namely construct that package known as a cyst and keep eukaryotic cells alive for two years, not in an ultracold freezer but right there on a shelf in the lab or in a package at the post office.

To make things simple and familiar, suppose you could design human cells that produce a viral protein, an antigen to stimulate an immune reaction. And suppose the antigen is one that initiates an immune reaction not only to SARS-CoV-2, which by the time you read this will have evolved into still further variants, but also to the next zoonotic virus sweeping through the world's human population as well as the ones after that. At that point—were you able to do something a mealworm parasite does but on your own terms—your engineering begins. Scientists studying basic biology, dragged along by their curiosity, produce innovations; engineers convert those innovations into money and jobs. By the time this conversion happens, however, the scientists, still driven by what their research subjects are teaching them, may have forgotten all about what they accomplished because some other parasitological problem has infected their minds.

Except for parasites like those inside these gregarine cysts, eukaryotic cells generally do not survive for two years sitting on a laboratory shelf only to wake up and begin functioning when, after all that time in a room, not in a deep freeze or packed in dry ice, they encounter whatever sends the activation signal. Humans are made of eukaryotic cells, those that have a nucleus contained within an envelope made from membranes, and the molecules that those cells produce, for example, hair, fingernails, muscle, nerves, and antibodies that fight disease-causing bacteria and viruses. All our domestic animals are also made of eukaryotic cells, as are our plants and fungi of agricultural importance. This fundamental property—cell nuclei with a membranous envelope—we also share with the infective stages inside the cysts produced by mealworm parasites.

Although some parasite eggs, especially those of roundworms, are also notorious for their viability over extended periods of time, even years, the conditions for such survival are typically warmth and moisture instead of a shelf in the lab.[16] Virtually all the global research on worm eggs is aimed at discovering ways to kill them instead of using them as delivery vehicles. With gregarine cysts from mealworm parasites, however, the targets for our imagined technology are exceedingly common, readily available, and in need of help globally. Those targets include not only us but also our livestock, for example, the 50 billion chickens a year we consume, as well as the 1.5 billion pigs, a similar number of goats and cows, and half a billion sheep you read about in chapter 8, all of which use veterinary pharmaceuticals.

The question now becomes: How many different kinds of cells can I put into these artificial cysts, the ones I've now made with the two-years-on-the-shelf property? That question, translated, means: For how many different diseases or genetic disabilities can I deliver the correction with this little package that parasites from mealworms made me think I should try to make? In this scenario, the question that would *not* need to be answered is one that plagues public health professionals globally and is the one that drives this discussion of mealworm parasite technology: How can I deliver this cure, this prevention, this vaccine, to the far reaches of our planet easily, quickly, cheaply, and efficiently? If you could do what those mealworm parasites do, you could ship the stuff anywhere in the world, without an ultracold freezer or dry ice chest, and put it on the shelf until you need it a year or two from now. What you can't do is what the parasites do regularly: make a cyst wall that functions to let you deliver your own package under those criteria, one that you, instead of a parasite in a beetle, have made. Nor, at this point, can we predict what that "use" might be; the walls of these minuscule cysts are a solution waiting for a problem, an incipient technology just begging not only for an application but also for a human being to invent, discover, or develop one.

How far-fetched is this bar talk about exploiting parasites for profit? Not as far-fetched as you might think. At least one member of another group of cyst-producing parasites, the Microsporidia, has been commercialized and explored as a means of controlling locust outbreaks in Africa and Argentina with mixed success. The species involved is *Antonospora locustae* (also known as *Paranosema locustae*), which naturally occurs in locusts (grasshoppers, to most of us).[17] If you have a grasshopper problem, companies with names like Grasshopper Attack will sell you billions of these cysts, also called spores, produced by infecting young grasshoppers in the lab. Do the spore bombs work? Sort of, but the spores need to be mixed with bait, and grasshoppers are notoriously erratic in their population explosions, so management of the pesticide is a bit of a problem.[18]

The cyst-producing parasites of chickens, however, have been much more cooperative as participants in commercial enterprises, mainly by helping make chicks become immune to more serious infections with the same parasite. The main culprit is named *Eimeria tenella*, a coccidian parasite that invades and sometimes destroys the intestinal lining, but there are several other species of the genus *Eimeria* that infect chickens and are capable of wreaking havoc on flocks of young birds. Typical vaccination technique involves infecting young chickens with small doses or attenuated strains that have been selected for their relatively nonpathogenic effect, after which the birds are relatively immune.[19] Commercially, the vaccines consisting of cysts, sometimes delivered in pellets that chicks can peck on, which gives them a mild, immunizing infection, are produced by infected chickens. Coccidian parasitism in poultry is an economic issue, given that we humans kill and eat about fifty billion chickens a year. So we have models for use of parasite cysts in commercial enterprises; nothing you've read so far in this chapter seems impossible, only extremely difficult.

The big takeaway from this discussion of parasite spores and cysts, those of gregarines in particular, is one we know well but seem to forget during stressful political times, namely that imagination is the source of so much of what we humans value: art, music, literature, and yes, scientific discoveries. A behavior that attempts to answer "what-if" and "why not" questions cannot be taught in school the same way we teach athletic skills, for example. Instead, that behavior must be fertilized by diversity, breadth of knowledge, the experiences of history, tolerance of strange ideas, encouragement, and confidence that anyone on Earth, any of our eight billion fellow humans, can come up with an idea that sweeps through various cultures like an ornery virus and generates efforts to bring that idea into fruition. This lesson could easily be taught by many different plants or animals, but parasites in mealworms turn out to be especially good at it.

I can envision John Steinbeck and his marine biologist friend and mentor Ed Ricketts, who collaborated with Steinbeck on *Sea of Cortez*, sitting on orange crates and having a conversation about these cysts and their economic potential. Their talk might quickly evolve into one about vaccination as a metaphor. One of them might ask: What kind of cyst might we invent that could carry a vaccine against willful ignorance? If this conversation had taken place about the same time as their original one, the answer would likely have been "books" and "stories." If it took place today, the answer could include "books," but then there would be a long silence. Both these learned and literary men would be considering the means internet "cysts," for example—we now have for delivering not only vaccines against willful ignorance but also cultural germs that seem to mutate so much faster than those vaccine equivalents can deliver an immunizing shot in the arm or to the brain. They'd probably shake their heads and order another beer.

CHAPTER TEN

MULTIPLE-KIND LOTTERIES

Mice, Parasites, and the Origin of Diversity

They all ran after the farmer's wife,
Who cut off their tails with a carving knife.

from "Three Blind Mice"; nursery rhyme[1]

The Glass Bead Game (*Das Glasperlenspiel*)

Hermann Hesse, 1943[2]

Decades ago, a young woman asked a question about mice. She had traveled into the western high plains to study nature and ended up in a group of other young people, all of whom had decided that "nature" translated into "parasites" and the "western high plains" was a university's biological field station with the mission of teaching and doing research on plants, animals, and microbes as they existed in the wild. All these young people had to choose both a project and a partner to work with them on this project. None of them had ever done such a thing before—choose their own assignment, then having chosen it, try to actually do it far from home, with few resources beyond their wits, some field equipment, dissecting tools, chemicals, and glassware, with only five weeks to accomplish the task and constrained by a seemingly arcane subject: parasitism.

This young woman had never killed a mouse before in her life either, but in wandering around the area, shooing mice out of her cabin and listening to talk about mice in other cabins, she wondered whether they had parasites. Then, perhaps as a result of talking to others about mice and parasites, she asked her scientific question: Do mice that live in the woods by the gate have the same kinds and the same numbers of parasites as the mice that live up in the hills near my cabin, those mice that seem to want to come inside and live under my bed? This question was, of course, a mouse version of the timeless one: Does where you live determine not only the infectious agents you encounter but also how and when you encounter them? The answer to this question, of course, is yes, but that answer does not reveal the details, for example, ecological factors that influence rates of encounter, numbers of parasite species involved, and the actual species, some of which are likely to have complex life cycles.

These details also remind us that we know more about ourselves—who might infect us with the idea that library books are more dangerous than assault rifles, what transmission mechanisms expose us to racial biases, and who spreads the belief that drag queens are a menace to society—than we know about lice, ticks, and worms. This young woman may not have had any idea about the kinds of parasites she'd find in and on those mice, if she found any. But what she did understand clearly was that her chosen question would lead her into the intellectual realm of scientists, a realm populated by people who answer to their own curiosity instead of their checkbook and who seem to see timeless questions in the most mundane circumstances, like field mice in your cabin, in the woods, and in the hills. She also understood that to answer her question, she'd have to behave like a scientist and dig through some mouse guts, blood, and fur—for worms, protists, fleas, and lice.

People who see a mouse in their basement usually do one of two things: call an exterminator or go to the store and buy some traps or poison or both, depending on how comfortable they are with the killing of small rodents. Depending on our killer/homeowner's background and education, the impact of such a decision can range from "Good, kill 'em and get 'em the hell out of my house!" to "Oh God, what do I do now?" From having dispatched a lot of mice during my own career as a parasitologist, helped others dispatch quite a few more, and having the mouse killers and dissectors I've trained go to work in the field and lab, the mouse-in-the-house scenario is pretty easy to imagine. You go to the hardware store and buy some snap traps, the kind that instantaneously crush a mouse's back or neck when the unsuspecting rodent follows its nose toward the peanut butter you've loaded on a trigger. Then you try to set that trap and place it where you believe mice run, all without it going off and smashing your finger.

Sometime in the night, from off in the dark, you hear a *snap!* Carefully, not knowing exactly what to do, you turn on some lights and check your traps. Sure enough, one is successful. You've never seen a mouse this still, this close, this freshly killed, blood draining from the nose. Having killed something, the question now becomes: What do I do with the body? If you're like most residents of the United States, you put on some rubber gloves, pick up the trap, careful not to touch the mouse or its blood, drop the whole mess into a plastic bag, drop the plastic bag into your garbage, and wheel the garbage bin out to the curb even though it's in the middle of the night and it will be three more days before the garbage truck comes by to empty it. But if you're a parasitologist, you put it in a plastic bag, which you then put in your freezer, company for the half a dozen dead birds that have flown into your living room window over the past year and are saved, yes, for their worms. Now you know exactly what this young woman will end up doing with mice in her basement a couple of decades after she sets her first traps in the hills behind her cabin and discovers some infections.

Five weeks later, having captured and dug through both the hair and the intestines of quite a few mice, she had an answer to her original question, wrote her project report, and summarized her results with a graph showing the numbers of mice with no parasites, one kind of parasite, two kinds of parasites, etc., in the two locations—hills versus woods. An example of that kind of graph follows. These sorts of pictures are not very difficult to understand; they're typical of what one sees on social media or in a local newspaper describing finances, crimes, or the progression of a pandemic. The young woman's original report has long been lost, so the actual numbers of mice (height of the bars) are fictitious, but the general results are not, and the numbers I've used to illustrate her results are pretty close to the maximum numbers of mice you could catch and examine for worms, fleas, lice, and ticks and also roughly identify the parasites captured during her allotted time. Nor is the idea that those results conveyed fictitious, and it's the idea that we're interested in, regardless of numbers. Her graph looked something like this:

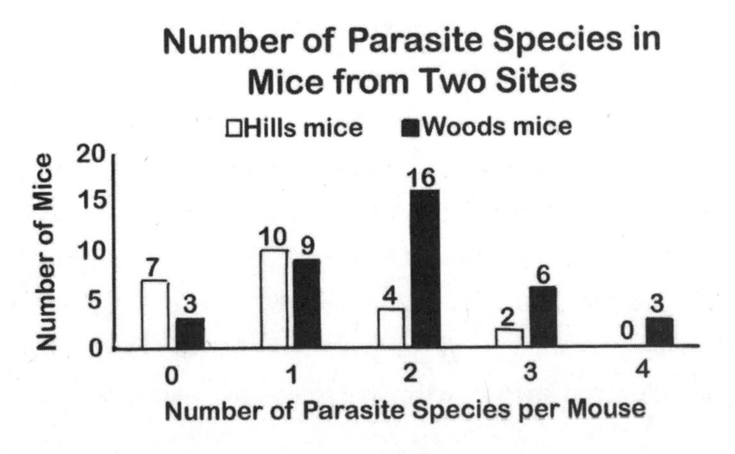

Now, what does this picture mean? What does it imply? What ideas does it spawn in a scientist's mind? And why is it connected to that political hot-button word "diversity"? What is important about this picture is not the actual numbers but that the hills are *different* from the woods, and not only can you see that difference in an instant, but it's also very easy to do a statistical analysis to demonstrate that difference to the satisfaction of fellow scientists. For example, up in the hills, there are no mice with four different kinds of parasites, at least among the mice this young woman collected while doing her research. And what's important about this picture is that the numbers along the bottom refer to the number of *kinds* of parasites, regardless of what kinds were represented. In other words, this young woman was looking at the results of multiple-kind lotteries, one going on up in the hills and another taking place down in the woods.

When she recognized this fact and told everyone about it, her recognition set off a struggle with computer programs, the goal of which was to develop a method of using multiple-kind lottery results to reveal whether members of a parasite community were influencing one another, an important question for parts of the world where humans are commonly infected with more than one kind of parasite. Her results had thus set off a major discussion among people who were with her that summer about how communities of parasites are assembled and organized and about the forces in nature that function to produce groups of species living in a particular habitat. Eventually that discussion became a contest between local parasitologists, with a case of beer at stake, over who could first write the computer program that would let us do the desired analyses.

Why did this discussion and the programming contest happen? Because the question of how human communities are assembled and organized is one that drives politics and religion, and parasitologists are particularly prone to transforming bar talk into what they call "big talk"—from worms to people within seconds. Such transformations are a form of intellectual play but with serious implications—serious, at least, for the scientists talking, because their exchanged and conversationally developed ideas infect their minds, never letting them forget about those global forces dividing, binding, and having an impact on their own local lives. It was simply too easy then, as it is today, to generalize this concept of chance—year and place of your birth and who your parents were—as major determinants of your future, your chances for a productive and meaningful life, and that of your children. But the software envisioned would allow analysis of communities far larger and more diverse than those living in and on mice, thus producing a research tool widely applicable to questions of how groups are assembled and sustained and in the process turning bar talk into something useful.

In the case of parasites, a community is a group of species and individuals of those species living in a particular habitat—in this case, a mouse, a single mouse, or a population of mice, depending on how you want to consider it. One tapeworm, one roundworm, three fleas, and two lice, in and on a single mouse, is a community of four kinds and seven individuals. Hundreds of tapeworms, roundworms, fleas, and lice in the hundreds of mice in the woods by the gate are also a community, with members linked mainly by their transmission mechanisms and the fact that they are living in a three-acre patch of woods in western Nebraska or on a rock-strewn hundred-acre hillside two miles away. The number of potential interactions between all these participants approaches infinity, especially when all the seasonal weather forces are considered. It should come as no surprise, especially to anyone who knows a parasitologist, that in the years following that young woman's destruction of mice and the subsequent production of that bar graph, the computer programming bet evolved from talk about a case of beer to talk about how to best understand how groups of species are assembled and subsequently controlled. The reason for this evolution is displayed prominently in newspapers and nightly television broadcasts, and it involves people instead of worms.

The mathematical techniques you use to describe the parasites in one mouse differ, as you might expect, from those used to analyze the hundreds in a population of mice. Those kinds of differences produce discussions and often disagreements about what forces shape and sustain communities. Do lice and fleas compete for mouse blood and tissue? Do the tapeworm and roundworm compete with one another for the food this mouse eats and digests for them? Or do these different parasites and individual fleas, lice, and worms basically ignore one another because that mouse's body is such a rich supply of blood and digested food, and you can't really compete for resources that are in excess? This student's graph suggests the answer to this last question is yes. That is not the answer most people who study communities, especially human communities, want to hear. People who study communities want to know about competition and dependence; they don't want to hear someone tell them that there are enough customers wanting pizza that two pizza parlors can exist a block apart and do just fine.

What this young woman had done to those who read her work and entered the computer programming competition was uncover, releasing it as an infectious agent, usually latent but now relapsing, a question embedded in every scientist's mind: To what extent does my internal logic system dictate how I view the natural world? At one end of this spectrum of worldviews is the assumption that any organisms living in the same environment compete with one another, so the most important questions for scientists are how they compete and for what they are competing. At the other end of this spectrum is the assumption that kinds of organisms do not compete with one another, and even individuals of the same species may not, indeed probably do not, compete with one another for available resources, especially if the resources are in abundance, like digested food in a mouse's intestine, so the really important questions are those surrounding transmission.[3] Put a human face on a worm and suddenly your worldview spawns motives, assumptions about the reasons why people are behaving as they do. As every parasitologist knows, those assumptions about motives can easily drive conflict, including killing, not so much for real worms as for ideas that behave like worms, infecting brains and altering behavior.

Put a worm face on the human, however, and the picture changes. The mouse may represent an abundance of resources, if not an overabundance, for a tapeworm, but getting to those resources is a matter of chance over which the worm has little if any control other than an evolutionary history, and evolutionary histories tend to be constraints rather than empowerments. Am I "better" in some way than those who are less successful in life? Is that homeless person lying on a grate in winter there only because he's lazy? That long-ago student project in which a multiple-kind lottery graph was produced will not go away. The concept of chance as a driver of fortune—a tapeworm in a mouse—becomes a stress-producing meme, perhaps best explained by an excerpt from one of my previous books, the subject of which is finally figuring out who my parents were from going through all the stuff they left behind after having died young:[4]

Blond + healthy + male + college-educated + with advanced degrees + between wars + natural-born United States citizen = a nice toss of the genetic, historical, and economic dice. I did not earn these traits; I was handed them. They are not a right; they are a gift from two people I never met until nearly a half-century after their untimely deaths, namely Bernice Locke and John Janovy. If there is anything that I have learned about the world since these traits were bestowed, it is that there are plenty of other kinds of gifts that humans receive upon conception—at the very instant sperm and egg unite—and that mine are neither exclusive nor, in some places, even desirable.

There are certain streets I cannot walk because of my unearned gifts bestowed by these scarcely known parents, and the same could easily be said for billions of others with whom I share Earth's resources. What I consider a gift, many others consider a curse, and if I decided to join some of those others in a social gathering, they might well kill me immediately. People everywhere are lucky if, through no fault of their own, they end up in the majority that possesses social and political power. The understanding of this principle is also a fortunate outcome of the education I received at the hands of those strangers Bernice and John, the public schools they supported with their taxes, and the university to which they sent me.

Thus, I ended up a *kind* through no fault of my own. I read my morning paper, and that word *kind* invades paragraph after paragraph, regardless of the subject matter, although most of that subject matter concerns crime, money, human behavior, and power. *Kinds* are doing something to other *kinds*, or trying to, and whatever they're doing depends on properties that none of them acquired through hard work but instead acquired by chance alone: skin color, hair color, place of birth, sexual orientation, parental resources, educational opportunities, mental capacities, languages spoken, ability to carry out certain enzymatic reactions, blood type... The list is fairly long.

That list of properties also gets played out against the ecological setting in which they appear. So just like a lucky tapeworm, I get up in the morning and enjoy the results of having found my way into an intestine, otherwise known as a university, overstocked with intellectual resources, surrounded by eager young minds, all of us soaking up arts and sciences produced through the ages. Yes, there are times when I imagine myself a tapeworm, immersed in everything humanity has done, said, and learned about the universe and the organisms that live in it. And as I walk down campus sidewalks, it's obvious that I'm a member of a community or an assemblage, depending on how you perceive the components—"community" inferring interactions of dependency and competition, "assemblage" inferring no such interactions.[5] Is there enough knowledge about the world that I don't have to interact with some math prof to live a comfortable life as a parasitologist in academia? Of course there is, but that friend in the math department can easily become one of the resources upon which my career is parasitic.

The word *kinds* is so critical to this discussion, to the big ideas that follow, and to the explanation for why people didn't want to hear this student's answer to parasite community organization, that we need a story—an analogy—to illustrate what was happening to those mice in the hills and woods. I've chosen your local library because libraries seem to be infective environments, although the infective agents are ideas and bits of knowledge instead of worms and lice.

Now, imagine that you are acting exactly like these mice but going to the library with an empty book bag. You walk into the building, hang around a little bit, doing what you typically do in a library, then leave. You are one of the mice in the hills, and you walk out onto the street, into this student's trap, with three books. You have no idea what those books are or how you got them. They just appeared in your bag. When you get home, you discover that one is about World War II, one is a biography of a country music singer, and one is a guide to flower arrangement. If your friends saw you with these books, stacked on your table in a pre-pandemic Starbucks, and one was open so that you seemed to be reading it, they'd believe you had impossibly eclectic tastes.

Frustrated with what you find in your book bag, you call up a friend whom you saw in the library that same day and ask her what books she checked out. She answers that she doesn't know, but she looks in her bag and finds one about antique cars, one about computers, a romance novel, and one full of Native American poetry, none of which she actually picked out. But both of you came home with books. You're the equivalent of a mouse from the hills who came home with three different kinds of parasites, and your friend is the woodland mouse who picked up four kinds. You are both like these mice, however, in that you have no knowledge of how these books ended up in your bags. You went to the library looking for a particular book, like a mouse searching for seeds, but came home with something unchosen and unexpected, like that mouse acquiring a flea or a worm.

Now, if you believe that I'm trying to teach you a lesson about infectious agents, you are correct. And if you end up reading those books and remember anything of what they contain, you have become infected by choice, although the worms are metaphorical—ideas, observations, interpretations, perspectives—instead of living, squirming, little animals wallowing around in your gut or clawing their way through your hair. Nevertheless, in this case, those metaphorical worms and lice could easily influence your behavior regardless of how you acquired them. Aside from their scratching fleas or mites, it's not always obvious that parasites influence the behavior of rodents, but most of us know from personal experience that idea infections can alter our own behavior. And that's why some of us go to the library.

If you have ever taken a statistics class, you are well aware of three classic models of chance and choice: the coin flip, the dice roll, and the bag of beads. If you have ever been to Las Vegas, you've used a model of chance and choice, and you probably realized you were doing so but didn't care because you were there to spend money on the chance that you'd win a whole bunch more money, knowing all the time that was not at all likely to happen. But this evening, at home, with little else to do, you decide to play mouse. You have a quarter, a pair of dice you found in a drawer, and a sack of beads of various colors, something left over from a long-ago craft class. You flip the quarter: heads. You shrug and roll the dice: two plus five is seven. You shrug again and decide you need some activity more challenging than a dice roll, so you reach into the sack of beads, retrieving a yellow one.

The important part of this story turns out to be the bag of beads, the stats class version of your library visit and mice in the woods. Because they are all the same size and shape and made from identical materials, blindfolded, you could never tell one color from another. Now it's time to play what my scientist colleagues have come to call the multiple-kind lottery. You invite some friends over for an evening of cocktails, fine food, fine wine, and games. Be careful with the wine; too much of it and the group may start assigning properties to beads of different colors. This game has the potential to alter the minds of your good friends, depending on what properties you all assign to the different colors and what you make of the results.

The friends who are coming over are your closest ones, those with whom you can talk politics and religion and never get in trouble. Now it's time to play multiple-kind lottery. You let your guests choose their role: hills mice or woods mice. You issue each person a notepad and pencil. You pass the bag of beads. Hill mice can choose up to three by reaching into the bag without looking but recording the results. Woods mice can do the same, except they get to choose up to four beads. Surely during this evening of frivolity, someone will mention Herman Hesse's book *The Glass Bead Game* because they were asked to read this book in college, learned that Hesse got in trouble with the Nazi government for writing it, and the morning news these days constantly reminds them of what their grandparents (or great-grandparents) said about World War II. But you're not living in Berlin in the late 1930s; you're living in the United States, and that ecological fact makes you wonder at the reading levels of Nazis who were able to decipher Hesse and apply his metaphors to them.[6] If all your guests were parasitologists, it might take them five seconds to assign colors to parasites: yellow, a species of tapeworm; red, a species of roundworm; blue, tick; purple, louse; green, flea; white, a second tapeworm species; and black, a second roundworm species. So what might be some rules? If you're a hill mouse, you can't accept one of the bead colors. And why not? Your explanation is your honors thesis. What do you need to know before you can explain why a hill mouse can't get the yellow bead tapeworm? You'll need to know how that species is transmitted; what kind of an insect must eat one of its eggs before a larva can develop into an infective stage; the ecological niche of that insect; a set of environmental factors, probably having to do with moisture, that restricts it to the woods; and how likely it is that a mouse will eat that infected insect, thus ingesting a young tapeworm that will grow into the constantly copulating, egg-spewing adult.

But what happens to this evening's conversation if we start assigning human properties to these colors? How much wine will it take before the beads evolve into different religious groups, different age groups, different sexual orientations, different skin colors, different levels of education? What happens to the conversation between you and your friends if you automatically return one color of bead to your bag, refusing to be associated with whatever that bead color has come to represent in your mind? You wake up the next morning unable to forget about how the previous night's game evolved, how a simple discussion of community relationships and your ability to decide them had an impact on what was basically a game of chance but one in which you imposed choices, the kinds of choices not available to woodland mice. That parlor game is an intellectual experience that could remain in your thoughts for a very long time.

Whatever became of the computer programming contest, the attempt to turn ideas unleashed by a young woman's mouse-killing, worm-collecting, and flea-gathering summer into serious science, published in some scholarly journal? Neither of the contestants won by generating a computer program that could be used to analyze communities with up to hundreds of members. Eventually, one of the contestants told his brother, an electrical engineer, about the contest. A few days later, a computer program, named THINKING, came delivered via email. The contest losers tried it; the program worked as expected so they published it, adding the brother's name to the author list.[7] But they never forgot his name for this page of computer code—THINKING. If you're going to try to analyze communities, you must THINK about the forces and factors that produce and sustain them, and you must also do something that tells you whether your thoughts are rational, logical, or derived from your imagination and desires.

Because of its relationship to the information age assault on our senses, the multiple-kind lottery could easily be the most commonly encountered yet underappreciated phenomenon that I have seen demonstrated by an undergraduate student's research project. In 2005, people in the United States sent about eighty-one billion text messages to one another; some of those messages contained infective ideas, even ones that could alter behavior—Let's do lunch—from which a recipient (host) selected a few, not all of them on purpose. Fifteen years later, the same people were sending and receiving well over two trillion such messages every year, including ones with pictures attached, again with the possibility of altering behavior—Thanks I'll send a card. Globally, that figure is now approximately sixteen million a minute.[8] But the question of how communities are assembled is one that applies to groups well beyond the parasites of mice, the text messages in your phone, or the deluge of visual ads that you picked up randomly just by watching television.

On a larger scale, the question of how communities are assembled and develop may be one of the leading questions of our time, given today's massive migrations and exploding human populations. Traditionally, at least in the United States, communities have formed and developed according to economic, ethnic, and geographic forces—money, race, and location. When those communities have changed in character, the change has typically been subtle until some event spurs action by some of the members. But I have this fantasy experiment in mind, one inspired by that parlor game with the beads, in which beads become real people that represent human diversity in all its forms. You're now the mayor of this town you've created by playing a multiple-kind lottery, drawing those beads at random, beads that represent the full range of human diversity—ethnicity, religion, talents, sexual orientation, physical traits, etc. Yes, I honestly believe that when this same group gets together for dinner and games, the results of this experiment will dominate table conversation. And from the players' attempts to manage such an eclectic mix to the benefit of all its members will come a substantial understanding of humanity's current status and behavior. That's quite a claim for the ultimate impact of one student's choice of a summer research project.

CHAPTER ELEVEN

IRON WHEELS

A Guiding Mythology Dismantled

> How one proceeds to put together a complete, coherent plot and a cast of characters all in need of development is as foreign to me as it would be for the house painter standing in front of that blank wall trying to figure out what to do next.
>
> Email from Joe Pollock to the author, 2015[1]

Every day, all around the world, on land, in lakes, and in the ocean, sharks, fish, frogs, salamanders, snakes, lizards, birds, and mammals are defecating into their environments and dumping tapeworm, roundworm, and fluke eggs along with their feces. If one of those eggs is lucky beyond description, it gets eaten by another animal of some kind—a beetle, maybe, or a snail or a small crustacean—and that other animal has exactly the right physiological properties to provide a perfect environment for that egg to hatch. If a tapeworm egg's luck holds out, the animal that has eaten it lives long enough for the hatchling to develop to an infective stage. And finally, with continued inordinate luck, that second animal gets eaten by a species like that from which our egg was originally passed. This immature worm then grows up into a reproductive machine, spewing its own untold thousands of eggs out into the environment through its host's feces. The same general pattern of life applies to roundworms and flukes too, although the hosts involved will differ, depending on the parasite, adding considerable complexity to what textbooks typically describe in relatively simple terms.

All this shedding, eating, growing, and egg production happens in the familiar straight line of birth, luck, life, and death—the luck not unlike that accompanying the birth and subsequent survival into old age of human beings born at the right time, in the right place, and to the right parents. But when a parasitologist describes a worm's development, for example while teaching it to a class of would-be physicians needing to understand how infections are transmitted, that teacher draws a circle, bending this line into what's called a life cycle diagram. Then that same teacher continues bending lines into circles with the lives of other worms—flukes, roundworms, thorny-headed worms—and single-celled organisms that invade blood cells when injected by a feeding mosquito and finally lice, bedbugs, mites, and ticks. Our teacher has been infected with a means of making sense of the world: life cycle diagrams.[2]

This practice of bending a sequence of events into a circle is illustrated by the following picture, showing how *Ascaris lumbricoides*, one of the most common parasites of humans on a global basis, is transmitted. This large roundworm produces thousands of eggs per day, which pass out into the environment with a person's feces (a). These eggs need to develop in moist soil before they become infective (b), (c), but upon ingestion by a person, the eggs hatch, immature worms migrate through the body, get into the lungs, are coughed up and swallowed, and mature into adults, sometimes by the dozens, in the intestine (d), (e). Nothing in these circular diagrams teaches us about the conditions of poverty and lack of proper sanitation that function to spread ascaris infections or about the impact these kinds of soil-transmitted parasites have on development of children in impoverished areas.

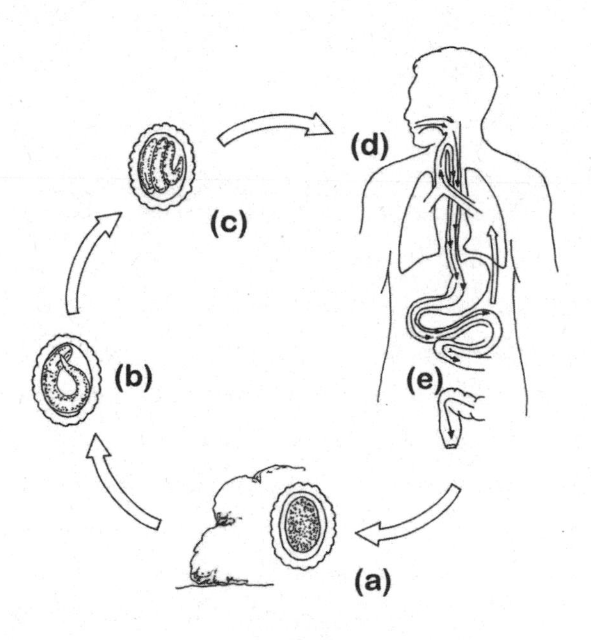

Out in nature, these individual parasites are living their own individual lives as straight lines from fertilized egg to death and decay, not unlike our human experience. If we show this life as a linear one, like in the following figure, it becomes easier to see time and space as factors influencing that life, the spread of worm eggs, and development into the infective stages. Linear representation helps us envision what happens to an individual as opposed to a kind of worm and thus helps a teacher explain how environments can alter the fate of individuals at any stage in their lives.

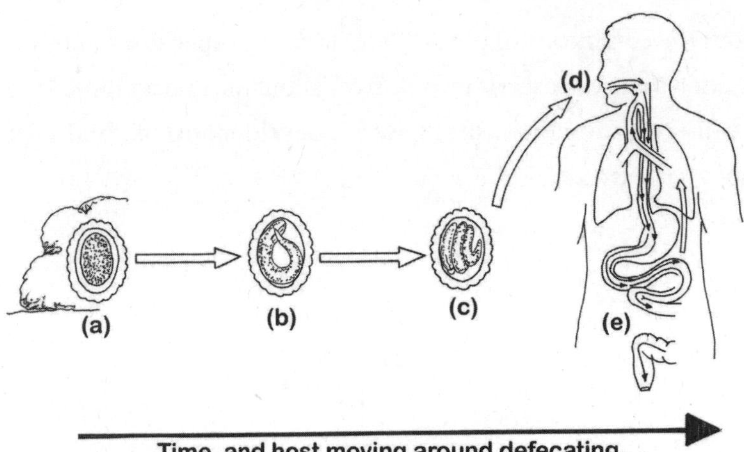

But what we scientists have typically done to simplify the complexity of parasitic relationships is focus on kinds instead of individuals. Although this focus is relatively harmless when all we're transmitting is information about how worms end up in frogs' lungs, the reduction of individuals into kinds is the ultimate political weapon of our times. Every one of our hot-button words—"immigration," "abortion," "race," for example—hides an individual experience, occasionally heroic but often if not typically tragic and rarely revealed unless some journalist tells us about it by separating the individual from its kind. Anyone who's actually tried to discover a transmission mechanism—for example, *Nematobothrium texomensis* in chapter 1 or *Distoichometra bufonis* in chapter 6—figures out in a hurry that all the important events happen at the individual level. No matter how much a teacher hides complexity in a textbook figure, the individual experience still rules in nature.

I suspect that if you ask one of those premed students a dozen years after leaving the classroom how hosts get infected, their answers will involve life cycle diagrams. These figures are almost synonymous with parasitology. No student has ever failed to encounter one on an exam; they are the rough equivalent of a society's defining myth. Parasite species life cycles—the scheme of things that we've built with our bending—becomes the way things are, a worldview as firmly constructed and seemingly indestructible as an iron wheel. The first time I heard this metaphor—the parasite life cycle as an iron wheel—it was at a national conference of the American Society of Parasitologists, where a young man named Matthew G. Bolek was delivering a talk to an audience of several hundred, mostly graduate students. He stood on the podium and claimed to have broken this wheel, then proceeded to tell us how he did it, what he discovered by doing it, and although he didn't state it explicitly, why his breaking was actually a dismantling of our guiding mythology, our means of understanding the complex world we entered when we started studying infectious agents.[3]

I'm telling you Matthew Bolek's story, or at least the part I know from my encounters with him, because with his research, he's shown not only that we're allowed to dismantle a scheme of things, that is, disassemble an iron wheel, but also that we need to do that before we can truly understand nature. Our life cycle picture is not necessarily the way things are, not the way we should be viewing this set of parasitological events; that's a lesson that goes far beyond parasitology. To arrive at this conclusion, Bolek dissected more animals while looking for worms than any other individual I know personally, animals ranging from the microscopic, in which case he had to design and build the instruments required to do the job, to larger ones, for example the largest bullfrog tadpole ever collected, at least up until the time he pulled it out of a stock tank pond in western Nebraska.[4] In my favorite photograph of Matthew Bolek, he's standing beside this pond and holding the tadpole, a veritable monster relative to other bullfrog tadpoles he's collected over the previous decade. All you can see are his hands; the world's largest bullfrog tadpole, almost nine inches long, is stretched out across both palms—a violation of your idea of what a tadpole should be. And that's just the start of your journey into the life of an infective worm.

This picture is a small reflection of what he told that audience during his iron wheel talk and what he told the rest of us over the next few years: *The world is not like you believe it to be; the world is not like you've been told that it is; the world is not like your textbooks are claiming it is*. In other words, with his research, he's telling us that the scheme of things is not necessarily the way things are, especially at the individual level. That bullfrog tadpole is only one of the participants in these endeavors; the parasites that will eventually live in its lungs, once it metamorphoses into a frog, are the main players. Bolek's arena is the global community of parasitologists and their students. At the time he was studying both the avenues for and constraints on movement of frog lung flukes through an ecosystem, he may not have realized how infectious his iron wheel metaphor was or how applicable to a nation invaded by a virus.[5]

History tells us myth replacement can happen when discoveries are so monumental that their impact on belief systems cannot be ignored, good examples being Nicolas Copernicus's demonstration that Earth was not the center of the universe and Charles Darwin's publication of *On the Origin of Species*.[6] What Bolek actually did was dismantle a prevailing scheme of things and replace it with an image best described as a baroque landscape. In his own words, he broke apart an iron wheel—the textbook picture about how parasites live in nature—and replaced it with a wreath, something resembling, intellectually, a holiday wreath made of evergreen boughs and sprigs of holly. In other words, he gave us something complex and fairly fragile, with lots of small options instead of hard and fixed, but still circular in basic design, although with an admission that the circle was a human construct, the bending of a somewhat linear path in order to fit an idea onto a page. The baroque quality is derived from evidence that the individual parts of this new construction may be interacting with their own environments independently of the eventual outcome, even though that interaction may be driving evolution of an entire system. The world is far more complex than your mind—that feature essential to your humanness but sometimes acting like a parasite itself—is making you believe by converting that world into an image you think you understand.

On a more important conceptual level, Matt Bolek is also telling us that what the authorities, for example textbook writers such as myself, are telling you about how the world operates is not necessarily a fact to be learned and repeated on an exam but rather something to be used as a guide to your own investigations. Although he started by breaking an iron wheel, he continues by showing us how to study, how to learn about what we've broken. Maybe "guide" is too strong a word, carrying as it does an underlying element of authority; perhaps "hint" would be more appropriate, or better yet, "clue." Authorities can hint without being authoritarian, but they can't guide without welding another iron wheel. Nature, the object of all this study, provides clues, but you have to be able to see them for what they are, thus training your eyes and mind to see the world differently than authorities claim it is. So maybe the best descriptor is a lens; through a lens, you can always see things that are hidden by the macroscopic view, which is always a simplification of nature. If your parasitic mind has any pathological effect on your behavior, it's to generate simplicity from complexity; that's probably also why politicians seem to be able to accomplish the same feat so easily.

Bolek's research on parasitism reversed that effect, making complexity from simplicity and in the process generating reality. His subjects were frog lung flukes. The common life cycle diagram shows frogs getting these worms by eating infected dragonflies. Immature dragonflies breathe by pumping water in and out of their rectums, which are modified to function like gills. The flukes spend time reproducing in snails before leaving the snail and getting sucked into a dragonfly's rectum. When the immature dragonfly matures into an adult and flies around, it gets eaten by a frog. The larval worms emerge, crawl up the frog's throat, and move down into its lungs, where they prosper and mate. The circular diagram of these events has been reproduced like a viral meme infecting textbooks and thus memorized by generations of premeds who will eventually encounter human diseases.[7]

Bolek started with a discovery by a previous student named Scott Snyder, who demonstrated that some species of frog lung flukes don't need to get sucked into a dragonfly's rectum, as shown in the common life cycle diagram, but can directly penetrate a large variety of tiny insects and crustaceans that commonly occur in ponds.[8] That demonstration, opening up a new way of looking at parasite transmission, was enough to win Snyder a national award given annually by parasitologists. Bolek followed up on this discovery by asking some simple questions: Who are these tiny insects and crustaceans—they are called second intermediate hosts, rather analogous to news sources that pick up stories, sometimes embellishing them, before transmitting them to another audience—and what kinds of frogs eat them? He then asked: What role do these tiny second intermediate hosts play in the movement of lung flukes through an ecosystem? That's the kind of question easily applied to hundreds, if not thousands, of infective agents, from viruses to nasty words to wacko ideas, that seem to move so readily through the human ecosystem.

What did Bolek discover that helped him break an iron wheel? First, that the movement of worms through the ecosystem depends mostly on which species is involved.[9] When dragonflies emerge from larvae into adults, most worms of one species embedded in their rectal gills were lost, shed with the rectal lining during metamorphosis. What the life cycle diagram showed as an essential step and what students accepted as a required developmental event was when most parasites of this species were eliminated from the ecosystem. Then he turned his attention to other species of frog lung flukes, those whose larval stages penetrated immature damselflies and other small aquatic invertebrates, ending up in all parts of the hosts' bodies. These species were not lost when their hosts metamorphosed and flew out into the nearby environment, and they could be transmitted if a tiny frog ate one of the small crustaceans containing an immature worm. Movement of infectious agents through an environment depended entirely on who these agents were, their evolutionary history, their ability to survive in certain hosts, and the local weather conditions. The iron wheel was simple; the real world was highly complex. Once you understood that complexity, you could start breaking your own iron wheels.

Our human realm, the cultural ecosystem in which we live, can be considered one gigantic exercise in the simplification of complexity. Single words—"gay," "evolution," "abortion," "socialism," "climate," "immigration," to suggest a few infective ones plucked from your daily news and social media feed—reduce truly massive phenomena, all with libraries' worth of historical baggage, down to single emotional reactions and votes. You fill in an oval beside a candidate's name because of that word "abortion," never giving a thought to prenatal care, early childhood development, economic conditions, educational opportunities, or the events—including rape—that might lead some young woman into a clinic. Nor can you talk about evolution around the dinner table at home, at least in some of the homes from which my former students came. And God help you if you're a teenager in an ultraconservative family and have decided that this evening's mealtime gathering is a good time to come out of the closet. What seemed so simple at the polls suddenly becomes amazingly complex at the local and individual levels.

The phrase "climate change" may be the most profound of these simplifications, given the impact that global warming will eventually have on weather, agriculture, coastal cities, distribution of diseases, and human health and migration. As of this writing, a query of the scientific literature database BIOSIS Citation Index, a resource available through virtually all American college and university libraries, using the search term "climate change review" produces 12,781 hits. Each of these hits is a scholarly publication, with topics ranging across almost every aspect of human existence as well as the fate of natural systems. A similar search of the EconLit database, also available through college and university libraries, yields 38,616 hits, the first one titled "Diverging Beliefs on Climate Change and Climate Policy: The Role of Political Orientation."[10] That title summarizes the problem of simplification: beliefs regarding global problems affecting all humanity should ideally be based on data and available information, not political orientation so often derived from a desire for power over people and nature.

The representation of nature, be it in writing or imagery, is really an invitation to begin your questioning of its truth, although perceived "truth"—the iron wheel Bolek used as a metaphor—in this case is not necessarily falsehood so much as it is an incomplete story. This relationship between representation and completeness applies to more than just worms that live in frogs. It doesn't take much mental effort to extend this metaphor and the methods for breaking it to political schemes, especially of those who would be kings. If we have learned anything about our own behavior since the turn of the millennium, it's that a collective mind can get sick from infectious ideas, so sick that it starts to imagine stories are true, and once that infection takes hold, it builds resistance to the mental immune system that we know humans possess. Bar talk among parasitologists could easily evolve into a discussion of books, guns, and the ideas that books, with their words' ability to invade our minds and emotions, are dangerous enough to be banned, while AR-15s, with their ability to destroy a classroom of children within seconds, are untouchable. The words "books" and "guns" hide worlds of complexity just screaming for us to understand them as they truly impact our lives.

How do we know that humans have the genetic capacity to resist the simplifying infection? The answer is because Mozart, Picasso, Beethoven, and Jackson Pollock were all born and extended the limits of our creativity. Members of our species *Homo sapiens*, literally the "wise man," are not necessarily constrained to believing what others tell us, especially if we are able to observe the contrary. We're not all Mozarts or Picassos, but neither must we behave, mentally and collectively, like simpleton sheep. We are individuals with power over our thoughts and deeds; every artist reminds us of that fact. Matthew

Bolek just went into the field to study parasites and then into the lab to confirm what he observed in the field, and he used that power with flukes that live in the lungs of frogs.

With his research on parasites, Bolek is also telling us, in very specific terms that are easily translated into general terms, the same thing that allegory writers have told us for a long time, namely that *how* you get from one place to another—the events you experience and the manner in which you experience them—is an integral part of the journey. He's also telling us that this particular journey is probably nothing at all like you tell your friends. Nor is it likely to be what your friends believe it to be. The world is not the way we think it is. He's also telling us that the circumstances surrounding your steps, stops, and starts could not only be constraints, but they could also provide avenues and options for completing your journey or deciding that this journey should go somewhere other than its initial goal. College students changing majors are an obvious illustration of this scenario. Matthew Bolek is doing for parasitology what the allegory writers do for our sense of destination. The allegory writers address the human experience; Bolek builds both a picture of nature and a set of instructions for interpreting that picture about life among the frogs, worms, fleas, ticks, lice, and ideas.

Although we humans develop in stages just like frog lung worms or even chickens and all other familiar animals, our stages are not nearly as distinct as those of the worms or the two that a chicken experiences: egg and hatched bird. A chicken egg's environmental niche is the nest, warmth, and rotation provided by a parent. The hatched bird also lives in a succession of niches as it grows, but because those niches overlap in space and contents, the stages appear to humans as a continuum. In other words, for a chicken, the environment occupied by a chick isn't necessarily so distinct from that occupied by a hen and not nearly as distinct, for example, as the inside of a snail and the inside of a frog's lung, two of the places occupied by a worm that needs to reach a frog's lung before it matures.

But with detailed observations, we humans could define the various stages in a chicken's life in what we might view as discrete and even devise words to define those stages the same way we sometimes do with ourselves: ovum, zygote, fetus, neonate, terrible twos, preteen, teenager, young adult, parent, senior citizen, etc. We could even represent that sequence in a circle, although that circle would represent our species, not any particular individual, and it would probably not include senior citizen or death, those human life stages with which everyone is familiar. In your mind, that circle could easily become an iron wheel. This is how humans develop, you'd say; this is the life cycle of the human. And in your mind, that circle would likely become "the way things are." At that point, you'd have built your own iron wheel, maybe missing the fact that some humans never become parents and some life cycle stages, such as teenager, were not always recognized as such, depending on the time and place young humans lived. Your "the way things are" would be an incomplete picture of the way human beings live and what the diversity of those lives could teach us about ourselves.

A textbook about Earth's biology used in a class on an occupied planet in another galaxy might easily refer to the development of *Homo sapiens* as this kind of iron wheel. That section of their book would talk about the destructive nature of this species, its reproductive potential, its inclination to fight and kill, its anatomy, and the technology that released it temporarily from the vicissitudes of nature. The *H. sapiens* iron wheel might look something like this:

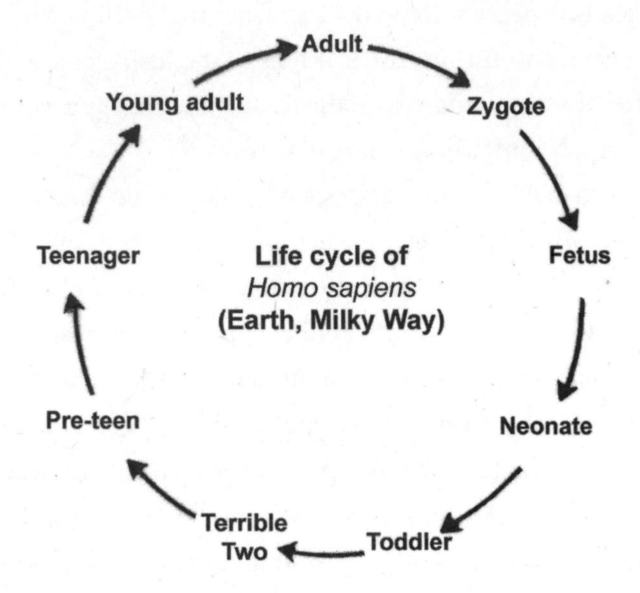

From our experience here on Earth, we can easily imagine a college biology class from outer space. Maybe these humanoids are on a planet named BS4722754 in the Andromeda galaxy, where our aliens have filed into a large auditorium to take a biology exam. The fourth item on a typical multiple-choice question might look something like this:

> 4. On Earth, a fetus develops into (a) a zygote, (b) an adult, (c) a terrible two, (d) a neonate, (e) any of these.

On Earth, most of the *H. sapiens* at a certain stage of development could answer this question correctly; in Andromeda, however, to assess the choices, students would need to have studied their vocabulary. The rest of the questions on this exam about *H. sapiens* would depend not only on what the visiting alien scientists had discovered after landing in Roswell, New Mexico, and returning to Andromeda after a few thousand years but also what the alien textbook editors had allowed in, given the space allotted to this species from out in the universe. What might these scientists have missed besides Mozart; Picasso; the Crusades; the NFL; piano lessons; Facebook; all of the hundreds of languages spoken throughout history; thousands, if not tens of thousands, of cultural practices; infectious diseases; the Olympics; the Vietnam War; Hiroshima; thousands of religions; slavery and its aftermath throughout history; and a virtually indescribable amount of information that could be assembled on this species? The answer to this question is whatever it would take to break the iron wheel constructed by those Andromedian textbook editors. Instead of a multiple-choice test—the quintessential simplification device—our alien students would need to get an essay test, then wander out into the night, look up at their sky, and wonder who else was out there and what they might be doing.

A mathematician named Douglas Hofstadter gave us another, more human and intellectual version of the iron wheel with his discussion of the relationship between the way things are and the scheme of things.[11] In this discussion, he reminds us that it is a common human practice, one an old biologist might consider an ingrained, even genetically determined trait, at least in many us, to make the scheme of things become the way things are. We interpret whatever we experience, be that experience physical, visual, auditory, or emotional, and that interpretation is an effort to make sense of what has happened to us. The scheme of things is also an attempt to make sense of our desires as related to our experiences. At one level—the physiological, biochemical, genetic, and emotional—we are exceedingly complex entities. At another level—the political, religious, that is, as a group—we are simple because we are simplifiers. Individuals, not committees, write novels; individuals, not legislatures, compose symphonies; individuals, not nations, solve differential equations and write software. Committees, legislatures, and nations reduce our complexity to a few rules and laws that we interpret as the scheme of things. Periodically we fight, including to the death, over our scheme of things having either become the way things are or driving us to believe that scheme must be made to be the way things are.

The other, truly major Bolek message is that evolutionary histories can be driven by forces that never show up in iron wheel diagrams, either the real ones or the metaphorical ones that humans tend to live by. Parasite lives are especially revealing in this regard because so many of them have several discrete stages and involve animals other than people. Worm egg survival can be influenced by one set of environmental conditions, but survival of a larval stage in some intermediate host is influenced by other conditions. Those thorny-headed worms in chapter 3 illustrate this fact because their eggs must survive on the open prairie long enough for a grasshopper to accidentally eat them, but once inside the grasshopper, that insect has to live long enough for the egg to hatch and develop into an infective larva. Selective forces acting on one life cycle stage are thus different from those acting on the next stage, but to survive, the parasite must adapt to both.

Those forces add layers of complexity to what seems on the surface to be simple processes. We're born, we grow up, get a job, get married, have kids, end up with grandkids, retire, write our wills, and die. That's our plan; that scenario is also the equivalent of an iron wheel textbook diagram. The big themes seem to drive human lives—births, deaths, money, sex, work, war—at least as presented in textbooks, especially historical texts. Through his study of worms that live inside frogs, Bolek is reminding us that the forces driving this trajectory can be and probably are much more diverse than we believe. And furthermore, those forces can be and probably are an integral part of our answer—both personal and collective—not only to the biological question, What is a parasite? but also to the ultimate question, What is a human being?

We humans worry about our jobs constantly. That worm in a frog's lung also has a job; everything and everyone has a job. But if you'll let them, the sounds your child makes while practicing the piano distance you—*Homo sapiens*—from that job and that worm. Those same sounds distance you from the political rhetoric pounded into an iron wheel, into that myth of who you are, what you are, what you should believe, who you should believe, and how you should behave. Those sounds produced by small fingers on a keyboard produce complexity. Accept them for what they are and enjoy them while they last. They are like the many environmental interactions of a tiny crustacean infected with a fluke destined to end up under a frog's tongue or some larval damselfly infected with worms designed for that same frog's lung when the damselfly gets eaten. Just like those piano notes add wonderful complexity to your own life, the conditions under which those crustaceans and insects live are really what shapes their lives, especially as individuals.

The life cycle diagram thus tells us a *fact* of life: There is a parasitic worm inside that damselfly, and when the frog eats it, that worm will crawl up the frog's gut and then make its way back down into the lungs where it will mature, mate with another worm, and start producing eggs. But water temperature, aquatic plants, rainstorms, wind, and fish and birds eating larval damselflies all make up the *form* of that parasite's life. Bolek tells us that the *form* of a parasite's life is just as important, probably more so, than the *fact* of that life. Yes, your child took piano lessons; that is a fact. The sounds of that same child practicing, however, represent the form of those lessons, and the memory of those sounds will forever transcend the fact that you wrote checks to a teacher. Yes, the contents of our minds as well as of our bodies come from diverse sources, diverse directions, and diverse origins. When we understand and accept these relationships, we've chipped away at our own iron wheels; when we don't understand and accept these relationships, our simplification genes have made our scheme of things become the way things are.

In a world where simplicity rules political rhetoric, often delivered by simpletons or at least by people acting that way, Bolek is telling us that if you want to understand how the world operates, you cannot accept simple and straightforward, and you especially cannot accept someone else's desire as a force driving your behavior. The iron wheels we build in our minds do not match reality, no matter how much we want that to be the case. But he's told us this lesson indirectly by showing us that our accepted—textbook—story of how the most common way of life among animals is maintained, out there in nature, is a simplified construction made by humans instead of the truth as lived by the species involved. It's not very difficult to extend that concept of a simplified story evolved into myth to virtually all human endeavors, especially those involving political power. What would today's American life be like if we, as a nation, were able to honestly disassemble the iron wheel that sometimes seems to be driving us into a fascist theocracy? That is a question worthy of an answer.

We humans are builders, but much of what we build is action based on ideas. We Americans are individualists; we succeed on our own skills, determination, and courage. We conquer wilderness; we are, above all, *good*. That is the iron wheel, polished by generations of politicians. But to those who've spent their lifetimes studying organisms that infect other organisms and reflecting on what that study has done not only to their analytical powers but also to their understanding of transmission mechanisms, the iron wheel comes to resemble a yoke, a massive necklace burden, not a device for rolling us into a future increasingly characterized by the infectious: words, ideas, rationales, technologies, and, of course, diseases.

The iron wheel metaphor is Bolek's contribution to the language of parasitology, and textbook writers who ignore this symbolism are missing a major opportunity to teach future scientists about how the world really works instead of how some previous textbook writer claimed it works. What is it about the discipline of parasitology that is so incredibly captivating, especially to a budding young scientist with no preconceived ideas about how the universe operates? The only legitimate way to answer this question convincingly is to collect something, cut it up, and find a worm, preferably a large one. Tapeworms seem to work best, although I don't know why. Because this whole discussion is now speculative, impressionistic, and totally without basis in scientific study or readily observed fact, I'll nevertheless offer an explanation. A tapeworm inside the intestine of a mouse you've just caught out in the woods, killed by separating its spinal cord, and cut open violates your sense of how the world should be organized. If you do the dissection with a reasonable amount of skill, the worm emerges from this mess of sliced tissue and partially digested seeds as a white band moving gently in the pan, almost undulating, not trying to escape but just being a tapeworm. The front end has a sort of searching, pulsing, kneading motion. You can easily imagine that "head" pushing itself into a gut lining, your gut lining. A textbook diagram comes alive.

This worm is physical evidence for an act of discovery. Pure, unadulterated, instant gratification discovery. It doesn't matter how many hundreds, perhaps thousands, of scientists and students have already cut up a mouse and found a worm. This particular worm is yours, your first one, alive, proof that something your mother used to tell you about cookie dough—"Don't eat that raw dough! You'll get worms!"—may in fact have been true. One tapeworm has altered your definition of "mouse." This cute little mammal that you remember so well from childhood stories and nursery rhymes is now a home for something else—a tapeworm. One story in particular has stuck with me as a result of reading it to our children—Leo Lionni's *Frederick*—mainly because of the art in our well-worn copy of the book and the expressions on the mouse faces. In the fields, Frederick's friends are all putting away grain for the winter, but Frederick is just sitting there, staring into space. His friends ask him what he's saving, and he claims it's the sunshine, the words that turn a bleak blizzard into literature. The other mice scoff at him. Winter comes; the mouse crowd asks Frederick for what he's saved, and here comes the poetry. In the story, the mouse crowd understands what it missed back in the summer.

What Frederick has done is save something inside him, something as alive as any idea, thought, emotion, memory—a particular way of looking at nature. His fellow mice are focused on their jobs or at least what they see as their jobs—get seeds, save seeds—or their lives—be born, get weaned, avoid owls, get seeds, eat seeds, find mate, have litter, avoid owls, die. Those linear lives could easily be bent into an iron wheel. They can't see that the poetry squirming around in Frederick's mind is just as alive as the tapeworm squirming around in his gut, but eventually they will find out what he has inside him. Winter comes; the mice are huddled, hungry, and cold; a great horned owl is hooting from a nearby oak. They ask: Where is the sunshine you saved for us, Frederick? And he delivers. The beauty of his poetry infects his community. The iron wheel of work, eat, starve, work, avoid owls, freeze has been broken.

We have a reason to study our lives the way Matthew Bolek studies worms. Such reflection can reveal not only a more elaborate existence than maybe does our constant preoccupation with day-to-day survival but also opportunities for change, for a kind of enrichment that celebrates our humanity by recognizing the value of our individual experiences and those of others with whom we interact.

CHAPTER TWELVE

JIGSAW PUZZLES

Missing Pieces

There are known knowns. These are things we know that we know. There are known unknowns. That is to say, there are things that we know we don't know. But there are also unknown unknowns. There are things we don't know we don't know.

Former U.S. Secretary of Defense Donald Rumsfeld, U.S. Department of Defense news briefing, February 12, 2002[1]

Scientists go about their daily lives knowing that they live on an island of understanding in a sea of ignorance about the subjects of their research, and that knowledge influences their perception of the world, their view of the daily news, and most of the actions they take as professionals. They also know that their work just makes this island larger, lengthening the shoreline as well as generating metaphorical landscapes throughout the interior—information valleys, lakes teeming with ideas, word streams, mountains of data, places where all kinds of problems are hiding, just waiting for more scientists to show up with their insatiable curiosity, binoculars, microscopes, and other observational tools. And in a manner consistent with the famous MacArthur and Wilson theory of island biogeography, the larger that island, the more complex its internal environment and the more diverse the occupying fauna and flora—although on an island of understanding, plants and animals are also metaphorical regardless of how wild they can be in some scientist's mind.[2]

As the island's research-driven shoreline lengthens, the boundary between understanding and ignorance also expands, providing more opportunities for scientists to do research on interesting problems. In essence, by trying to solve problems and answer questions, scientists generate more problems and questions. Thus, one could argue that the main product of science is known unknowns, that is, an awareness of our ignorance. By attacking the known unknowns, scientists tend to produce more known unknowns as well as remind themselves that there are plenty of unknown unknowns. And among scientists, those who study animal parasites may be the most notorious for uncovering what we don't know and understand about life on Earth.

Why are parasitologists so good at discovering gaps in our knowledge? The answer is mainly because there is so much parasitism to study and it manifests itself in such diversity throughout the animal kingdom. That diversity is on display in every textbook that includes information about parasitic protistans, worms, ticks, mites, and lice. Parasite lives are often chancy and complex, highly adapted to their hosts, subject to environmental vagaries, and locked into developmental sequences that may require them to encounter strikingly different habitats in a limited time.[3] For this reason, a simple question can unleash a career. For example, parasitologists facing biology's most persistent question—What is it?—can easily be dragged into libraries filled with foreign-language literature, museum collections from around the world, and molecular laboratories that open windows on never-imagined evolutionary relationships, upending traditional wisdom and showing us that appearance can be remarkably deceiving—a lesson we may already have learned from observing our friends but that now applies to much of the natural world. A certain kind of mind just cannot resist spending its life in such rich intellectual environments, standing on the shore of that metaphorical island, looking out over the ocean of ignorance, and deciding to build a boat.

Unlike ornithologists, for example, who are just as driven to satisfy their curiosity but might focus on the life of a single bird species or a group of related ones, any parasitologist who studies parasites of birds must address questions not only about the birds and how their biology affects their chances of getting infected but also about all the other organisms that live in and on them, as well as how those parasites got there, whether they are restricted to that bird species or occur on either related or unrelated hosts, and whether the evolution of those parasites maps onto that of their bird hosts. The same can be said of any parasitologist and their chosen host targets, be they mammals, amphibians, reptiles, fish, sharks, or the world's invertebrates; and there are a *lot* of invertebrates with parasites.

These animal hosts also live in their characteristic ecological settings, and they carry out their life functions in specific ways—feeding and reproduction, mainly—since they evolved their specific traits, with no consideration whatsoever for human activities such as war, political strife, and habitat destruction. The same can be said for their parasites, although with complexity generated by multiple life cycle stages and the different ecological requirements involved, that is, everything added to whatever comes into a parasitologist's mind when reading one scientific name. And if a host species is invasive, having ended up in a part of the world different from where it evolved, it can sometimes carry its parasites with it and share them with native fauna, increasing the complexity of a system some parasitologist is trying to understand.

When studied by scientists, complexity breeds awareness of what we don't know or understand, that is, recognized and usually admitted ignorance. And like the case with political rhetoric in which one word can compress history into a simple emotional response—think "slavery"—infectious agents teach us that complexity associated with parasitism cannot truly be simplified or compressed into a single word. "Malaria," "toxoplasmosis," "filariasis," "acariasis," and "schistosomiasis" are all just words, yet each one conveys a massively complex universe of disease, pathology, immunity, economics, history, demographics, cultural factors associated with treatment and prevention, political consequences, and evolution, as a minimum.[4] The same could be said of today's hot-button words, although you might never know that from media coverage of current issues or your neighbor's yard signs before an election. Maybe to illustrate the relationship between knowledge, complexity, and ignorance, we need a concrete example, a lesson from parasites and a parasitologist, so I'll pick one provided by a former student—the jigsaw puzzle.

To appreciate why a jigsaw puzzle is such an effective image of what we don't know about life on Earth, I need to introduce you to this aforementioned student. Back in the 1980s, I was teaching a college course named Zoology 112 with an enrollment of about two hundred, and I was responsible for both lecture and lab. One of those students, Richard Clopton, majoring in entomology at the time, told me that the insect specimens we'd been using as instructional material in zoology lab were worthless. He claimed, correctly it turned out, that they were in such poor condition students couldn't see the important structural features and thus couldn't really learn about insect diversity at the introductory level. That was the first and still the only time an undergraduate student has delivered such an evaluation of my teaching and done it in person.

How did I respond to this claim, this challenge? At the time, I was also director of the Cedar Point Biological Station, a mini campus in western Nebraska that hosted advanced field courses, so I offered Clopton ten weeks of free room and board if he'd spend that time making an insect collection, one to be used in teaching, that met his standards. He accepted the offer, we talked about what this collection should include, we bought the supplies, and at the summer's end, he delivered an essential piece of the puzzle known to my students as "zoology lab." The specimens he collected and prepared with what are called Riker mounts were used for the next thirty years. Some of them are probably still being used; one of them, a favorite, sits on the shelf above my computer.

Richard ("Rich") Clopton is now a faculty member at Peru State College, a mainly undergraduate institution in eastern Nebraska, where he teaches a variety of biology courses, serves as editor of the *Journal of Parasitology*, and defies generally accepted American higher education logic by being not only a small-college prof with heavy teaching obligations but also the editor of a peer-reviewed scientific journal, an actively publishing scientist, and a world expert on what is probably the most diverse group of eukaryotic organisms on Earth, the single-celled gregarines, those Shmoo-like cells mentioned in chapter 9. Here's what they look like:

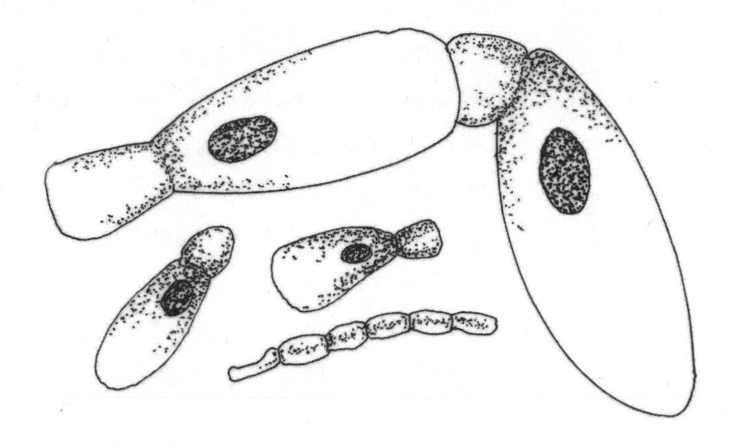

Why is this group of parasites the most diverse of eukaryotes? They parasitize arthropods, annelids, echinoderms, and several other phyla. Based on what we can estimate from Clopton's research, there are at least a million species of these parasites in the world's beetles alone, of which scientists have discovered fewer than two thousand. Even if he lives another hundred years, Clopton will never see, touch, or dissect all the world's beetles and their gregarines, no matter how hard he tries and how far he travels in his effort to understand the microscopic communities living in insect intestines. In casual conversation about these parasites, he could easily have been the author of that epigraph quote at the beginning of this chapter, attributed to Donald Rumsfeld.

Instead of quoting Rumsfeld, however, Rich Clopton is standing in a room beside a screen image of a thousand-piece jigsaw puzzle and doing what scientists do regularly—talk about their research. The occasion is an annual meeting of the Rocky Mountain Conference of Parasitologists, held annually at the Cedar Point Biological Station, eight miles north of Ogallala, in Keith County, Nebraska, the same place where, decades earlier as an undergraduate, he'd assembled that teaching collection for generations of other students he'd never meet.[5] But only one of these puzzle pieces has any color or pattern; the remainder are empty, and they represent all the known unknowns and unknown unknowns in only one group of parasites. His subject is the number of species about which we are ignorant, and that includes only the gregarines. If he'd been talking about other groups of organisms, there might have been fewer pieces in this puzzle, but never would all of them be filled to produce a complete picture. Every biologist knows that we'll never complete our inventory of life on Earth before we destroy it.

This puzzle on the screen has a thousand pieces because that's the image Clopton has chosen to illustrate his point, and it's a nice one for a small audience, never implying that we actually know 1/1000 of the species of parasites. He's giving an invited talk, the Gerald D. Schmidt Memorial Lecture, one that honors a deceased but highly respected parasitologist who discovered many new species of worms. When he points to the one piece with color, everyone understands the symbolism because his audience consists entirely of parasitologists, people who probably know better than any other group of scientists just how ignorant we are about what lives on our planet, including those parasite species hidden in or on other undiscovered host species that live on our planet.

His audience also symbolizes the history of a discipline, given that its members range from retired scientists marveling at what modern molecular technology can show us about evolutionary relationships to grad students and even some undergrads now using the most current technology to address age-old questions about how parasites live in nature and to whom they are related. But among these audience members are those who have described numerous new species, some worms, others of single-celled parasites such as Clopton's gregarines, and they're sitting in a place known for new species discovery. This is the very room, for example, where an undergraduate named Sarah Richardson worked after she came to the field station, telling her sister, "I'm going to find a new parasite and name it after *you*!" And she named that parasite, discovered in damselflies and described as a new species, *Actinocephalus carrilynnae*—an honorific, of sorts, to her sister Carri Lynn.[6]

That jigsaw puzzle image also reminds Clopton's audience of the efforts involved in fitting just one new piece into the picture, given that the one he's filled in represents two or three centuries of research by many people from around the world. Most of those audience members understand that his illustration of knowledge versus ignorance applies far beyond the field of parasitology, as Rumsfeld reminded us prior to our nation starting a war of our choosing and admitting that there was plenty we didn't know about the situation or the consequences. These audience members also know the main reason we'll never complete the puzzle: a significant fraction of Earth's biota is simply so inaccessible that it can never be studied in detail, and another significant fraction is being destroyed, on purpose, at a hellish rate, a good example being the clearing of Amazonian forests—over three hundred thousand square miles since the late 1970s.[7] Most if not all that biota is parasitized; their puzzle pieces will remain blank forever, as will the pieces of whatever groups their parasites represent. The sound of Brazilian chain saws is actually the sound of humans producing unknown unknowns.

In contrast to the vast number of species that are inaccessible to scientists, some others are remarkably easy to maintain in the lab, and as a result, those species tend to provide a disproportionate fraction of the picture we have of living organisms on Earth, good examples being the fruit fly, *Drosophila melanogaster*, and the near microscopic roundworm, *Caenorhabditis elegans*. We parasitologists call these latter types "cooperative"; not only are they inextricably linked to the humans who study them, but they also shape our perception of life on Earth, especially the functional aspects of that life, to a remarkable extent. For example, there are about 150,000 known species of flies in the insect order Diptera, to which *D. melanogaster* belongs, but that species is probably as famous as a single fly species can get because it has helped us learn how genetic traits, including some of those in humans, are inherited.[8]

Although there is plenty of research on other fly species, especially mosquitoes, fruit flies are cooperative enough that they are easily used in high school and undergraduate science classes. In the summer, they can often be captured with a slice of banana and kept as pets, although tolerant parents are also part of the culture medium. My own initial experience with *D. melanogaster* was an eye-opener because it demonstrated so clearly what you could learn about nature if some small piece of that nature would live with you instead of out in the jungle and, furthermore, perform like you wanted it to in the lab, predictably, on schedule, and happily, or so it seemed. This level of accessibility has two outcomes: it focuses our attention on interesting experiments, the ones we can actually do because our organisms are so cooperative, and for that reason, it shapes our sense of what the world is like while hiding how little we know about the rest of it. We invest so much time and intellectual energy in our relationship with one kind of organism that we tend to assume that species is representative of its group, whether it is or not. And regardless of how friendly is *D. melanogaster*, it is certainly not representative of flies in general, a group that includes mosquitoes plus another hundred thousand species. But I should also admit that those experiments can easily become so interesting that they consume our curiosity; that's what sometimes happens to scientists, especially to those who end up building their careers on that curiosity.

Caenorhabditis elegans plays somewhat the same role as fruit flies in shaping our knowledge of roundworms,[9] especially their genetics and development, but the economic importance of parasitic nematodes has also produced a massive body of research on other species. However, nothing you do with *C. elegans* in a lab teaches you about roundworm parasitism on a global scale. No high school biology teacher designs experiments using parasites of humans, for example, *Onchocerca volvulus*, the roundworm that causes river blindness; *Wucheraria bancrofti*, the causative agent of elephantiasis; *Haemonchus contortus*, which produces liver disease in sheep; or *Dirofilaria imitis*, dog heartworm.[10] That teacher may not know anything about those parasites causing human disease. But *C. elegans*'s cooperativeness is illustrated beautifully by a quote from Michele Lemons, a teacher at Assumption College in Worcester, Massachusetts: "This nematode is an ideal choice for experimentation in an undergraduate lab due to its powerful genetics, ease and low cost of maintenance, and amenability for undergraduate training."[11]

The experiments discussed involve nerve impulse transmission. Although these functions occur at the cellular level, *C. elegans* behaves in a way that lets students infer function. For example, anthelminthic drugs known to inhibit synaptic function paralyze the worms; an instructor can provide cultures with known but not revealed mutations and then ask students to use these drugs to determine where, among the worm's many genes, those mutations have occurred. The fact that *C. elegans*'s entire nervous system has already been mapped and its DNA completely sequenced allows this kind of study. Students can write hypotheses, test them, record and analyze data, and report the results—all typical scientist activities—economically and within the time available to high school and undergraduate students.

In an ideal world, however, a schoolteacher using *C. elegans* in a lab might also ask students to collect some roundworms from the pond in a nearby park, try to identify them, and keep them in culture to also use for experiments, maybe to compare a roundworm they don't know with the one they do know about. That exercise would likely end up as a tour through the internal anatomy of nematodes and an unforgettable lesson in frustration. But it would also be an introduction to what a person needs to know before even trying to answer that most persistent question in biology—What is it? That introduction, along with the question, has the potential to turn a high school student into a scientist.

Biologists estimate that in addition to *C. elegans* and the species of medical, veterinary, and agricultural importance, there are about a million species of roundworms, both parasitic and free-living ones, although only about thirty thousand have been officially described, another way of saying "discovered."[12] How easy is it to discover (describe) a new species of roundworm, especially one that you've gotten from a host species that might not have been studied very much? The answer is it's not very difficult if you have well-preserved specimens, the time, appropriate literature, laboratory resources such as high-quality microscopes to take digital photographs, the ability to draw elegant pictures of worm anatomy, sequencing machines to decipher DNA samples, computers to run software comparing your

DNA sequences with published ones, and patience with anonymous reviewers and editors who will ultimately decide whether your name and whatever name you have proposed for this worm will end up in the permanent record of life on Earth. If some biologists tell you it's easy to both discover a new species and publish its formal description in a peer-reviewed journal, they're lying.[13]

In the five years prior to the writing of this book, more than seven hundred scientific papers were published with descriptions of new roundworm species. The overwhelming majority of those papers concerned parasites; of those papers published in just the year prior to this chapter being written, the hosts and sources involved were, as a minimum, frogs, toads, salamanders, fishes, badgers, hares, beetles, millipedes, cockroaches, slugs, Black Sea sediments, tropical forests, coastal sand dunes, pine logs, figs, several species of mice, and soil from around plant roots.[14] There are a lot of scientists working on their chosen parts of that jigsaw puzzle, trying to find out what lives on Earth and adding to the inventory of known species. Ideally, nerve impulse transmission experiments with *C. elegans*, as suggested by Michele Lemons, generate enough curiosity about roundworms in general so that a few of those high school and college students decide to study nematodes in more depth, discovering parasites in the process. If that happens, and curiosity ends up driving a young person to spend the rest of their life trying to answer the fundamental question in parasitology—Who's infected with whom?—that budding scientist will discover a whole lot of missing puzzle pieces, in the process not only changing some unknowns into knowns but also some unknown unknowns into known unknowns, that is, generating ignorance by producing information.

In contrast to the more than seven hundred new roundworm

species described in the past five years, there were only eighty-three papers published on gregarine parasites—Rich Clopton's specialty—with fewer than fifty new species being discovered and formally described.[15] At this rate, if Clopton worked as hard and as successfully as all the rest of the world's scientists who study gregarine parasites combined, it would take him about twenty-five hundred years to discover all the ones in beetles alone. But he's never confined himself or his own students to studying only the parasites of beetles. Over the course of his career, he's explored the intestines of damselflies, grasshoppers, and roaches, regularly discovering new parasite species and giving them names, sometimes in honor of other scientists, sometimes based on their unique structural features, and sometimes reflecting the host in which they were discovered.[16]

None of these gregarine parasites cause human disease, debilitate domestic animals, or stunt the growth of corn and soybeans. For that reason, they are typically considered unimportant by much of the scientific community. Instead, as a group, they are accessible and beautiful under the microscope, have relatives throughout the world and throughout much of the animal kingdom, and provide a jigsaw puzzle of enormous dimensions. The range of hosts parasitized by gregarines suggests they originated, along with their parasitic habit, at least four hundred million years ago, evolving from single-celled ancestors that spawned many other descendants, including ones that now live freely in many habitats throughout the world. What gregarines lack in pathological effects, they more than make up for with their opportunities for intellectual engagement with the natural world, and that is what makes them far more important than most scientists realize.

Over the long term, when humanity seems desperate for

scientifically literate leaders, opportunities for meaningful study of nature at an early age are potentially as valuable as a new pesticide or drug to combat some disease. This last feature may be the most important life lesson delivered by a parasite: humans are curious almost beyond description, exploration is deeply embedded in our species' genetic makeup, and a vast puzzle is always alluring, infective, and consuming. Some of us get on a rocket to the moon, others send telescopes into space, but still others go looking for parasites. It doesn't take much effort to find some; it's the attempt to identify them and maybe recognize that what you've found is new to science that generates the work as well as the allure—that is, the infective idea. If, instead of stomping on a cockroach, a kid decides to carefully dissect it, maybe using pins, needles, and a razor blade, gregarine parasites are likely to pop out of the gut. Finding those large one-celled parasites also requires a self-taught lesson in insect anatomy, a fringe benefit. That's a lesson easily delivered to children old enough to hold a magnifying glass. The other life lesson is that there are as many missing puzzle pieces to be found under a microscope as there are at the end of a telescope or in a printout of nucleotide sequences in DNA.

What Clopton's parasites share with the more glamorous and economically important species, however, is a means of attaching to their hosts, at least during part of their lives, and as is the case with most other parasites that attach, the structures that function in this way can be elaborate. For example, among the most beautiful of all parasites are tapeworms that live in sharks, skates, and rays—those cartilaginous fishes, the so-called elasmobranchs, that have been swimming in our planet's oceans for at least two hundred and fifty million years. These worms attach to a shark's intestine by means of a scolex—the "front" end, the organ used in attaching to a host's intestinal wall—which depending on the species can be elaborate, with grooves, suckers, hooks, and tentacle-like structures. A scientist named Janine Caira from the University of Connecticut has devoted her career to the study of these tapeworms, generating new species descriptions, stunning electron microscope photographs, equally stunning ink drawings, evolutionary histories based both on structure and on DNA sequences, and graduate students who become professional scientists, continuing this general line of work.[17]

What have we learned from Caira's lifetime of research on the classification and evolution of tapeworms from elasmobranchs? We've learned the same thing Clopton is teaching us but with a different group of animals: a massive and incredibly rich body of information about our planet can be produced by the curiosity-driven activities of a single individual, especially when that person is in a position to collect not only parasites but also followers who are equally as curious. Caira obviously considers tapeworms from sharks to be vitally important regardless of their seemingly nonexistent impact on human health, economics, or political events; she shares that trait with Clopton and his focus on gregarine parasites. An exploration of scientific literature reveals this same situation in group after group of plants, animals, fungi, and especially microbes like viruses and bacteria. Time after time, scientists have told us just how ignorant we are of the organisms that live on Earth and warned us that such ignorance is not to our long-term benefit.

In addition to showing us what we never knew about worms that live in sharks, Janine Caira has also produced exactly what some of the most famous artists and writers have produced, namely a greatly expanded vision of what is possible and probable among living organisms on our planet. The electron micrographs of elasmobranch tapeworm scolices (plural of scolex) totally reshape whatever image most of us, including parasitologists who have handled tapeworms from local mammals, have of parasite attachment organs.[18] Although these worms are sometimes tiny, only an inch or so long, their scolices can be elaborate almost beyond description, with folds, cavities, and extensions. Furthermore, the surface of these worms can be complex, with spine-like projections seen only with a scanning electron microscope. Because we ask why those holdfast organs are so beautiful and complex, Janine Caira's images also conjure up all the potential interactions that occur between two unrelated species, from the molecular to the population and evolutionary levels. In this way, she shows us what other aspects of the host/parasite relationship we could and probably should be studying.

Similar complexity but at the cellular level can be seen in structures that some of Rich Clopton's gregarines use to hang on inside a damselfly's gut—membranous extensions of a cell membrane but shaped like hooks or tentacles. These structures are also downright beautiful under a microscope, but scientifically, they suggest a kind of host/parasite interaction that might be the basis for research on the evolutionary events that lead to such elaboration. But you can't breed damselflies in the lab the same way you can breed fruit flies, so you can't provide yourself with a steady supply of material upon which to conduct the necessary experiments. There are more than six thousand species of dragonflies and damselflies, informally called "odonates" because they comprise the insect order Odonata. All these species spend their larval lives as underwater predators before they metamorphose and become flying predators, consuming mosquitoes and other flying insects as rapidly as they can catch them. Both larval and adult odonates have gregarine parasites whose cysts must develop in water before becoming infective. The questions of how a flying insect gets cysts from the bottom of a pond and whether the same parasite species infect both larval and adult hosts add mystery to the structural beauty of a gregarine's anterior end.

Scientists Caira and Clopton can be thought of as soldiers in a war against ignorance about our planet, a war that many other scientists admit cannot be won, largely because of the speed with which we humans are destroying nature. The scientific community has been telling us for decades that we are wiping out the resources needed to survive long term, but as a species, we have not been listening. Paul Ehrlich's 1968 book *The Population Bomb* sounded a warning, and in January 2023, Ehrlich was interviewed on the CBS news show *60 Minutes* where he updated the viewing audience on the long-term impact of human population growth—more than doubling since 1960.[19] He's not alone in predicting a dire future for a species, *Homo sapiens*, that not only is running out of natural resources to support its population but also consuming those resources, largely fixed, at an astonishing rate. The world's scientific community recognizes this aspect of our existence but is largely stymied in its attempt to change our collective behavior. We're not only ignoring those blank puzzle pieces, but we're also throwing them away.

In every case, from gregarine parasites in beetles to thorny-headed worms, tapeworms, ticks, mites, fleas, and flukes, the intimate interactions of parasites add both known and unknown unknowns—missing pieces of the jigsaw puzzle that represents our understanding of how life truly exists on a single planet and how the pieces fit together. The big take-home lesson from all these parasites and their hosts is that although our island of understanding of what lives on

Earth and how it lives, surviving all the ecological vagaries our planet can deliver, is massive, it's nevertheless tiny compared to the sea of ignorance. The puzzle will never be assembled, although that fact will never deter scientists from trying to do so. There is plenty of evidence from the fossil record that animals have survived parasitic relationships or ones similar for hundreds of millions of years. Humanity's presence on this planet is but a fraction of that time. What's missing is that puzzle piece telling us how to survive as a species for as long as there have been beetles and gregarines sharing our common environment. Where might we find such a piece? Probably in some climate-change denier's garbage bin.

CHAPTER THIRTEEN

MANIPULATION

Obeying Your Parasites

The principles on which a firm parental authority may be established and maintained, without violence or anger, and the moral and mental capacities be promoted by methods in harmony with the structure and the characteristics of the juvenile mind.

Jacob Abbott, 1871[1]

Of all the effects that parasites have on their hosts, manipulation of host behavior seems to have an irresistible power to map itself onto the human psyche, possibly because of the symbolism involved. When an ant, infected by larval worms that will mature in sheep, climbs to the top of a grass blade, thus behaving in a way that guarantees a sheep will eat it and become infected with the adult worms, we immediately understand what is happening—a parasite has made its host vulnerable to predation and death. If ants could talk, the colony queen might ask one of her subjects why the dangerous behavior, and the answer would involve blame: The worm made me do it.[2] When mice infected by a single-celled parasite are asked a similar question of why they seem to lose their fear of cats, the answer would be the same: An infection made me do it.[3] And in the information age, when we ask the same kind of question about human behavior that sometimes seems counter to the public good, parasitologists can easily credit an infective idea, a meme, a powerful leader's rhetoric, or technology that intrudes upon our innocence. All these cultural germs, transmitted as larval ideas or by handheld vectors, are produced by people. We humans have no problems placing blame or understanding the metaphor.[4]

The worm that infects ants is a notorious example of parasites manipulating host behavior, and it has been studied, explained, and written about extensively to the point that the causative agent's name, *Dicrocoelium dendriticum*, shows up frequently in popular writing about parasites. Adult worms live in the bile ducts of sheep and cattle, causing fibrosis and thickening; the worms shed eggs, which pass out with the host's feces and are eaten by land snails. The resulting larval stages reproduce asexually and then burrow out of the snail, which produces mucus, wrapping the larvae in slime balls that get eaten by ants. Ants evidently love these slimy packages of worm larvae, and when eaten, most of the larvae end up in the ant's abdomen, but a couple of them migrate to the head and attach to a nerve ganglion, thus becoming "brain worms." These latter larvae are not infective, but they manipulate the ant to climb up grass blades, lock its jaws around the grass, and stay there during the day, virtually guaranteeing that it will be eaten by sheep or other ungulates, thus infecting the next host with those hundreds of larvae in the ant's abdomen.[5] Because those "brain worms" are not infective, we humans typically describe them as altruistic, guaranteed to die so that their siblings can have a chance to become adults packed into a bile duct, breeding like crazy and dumping eggs back into the pasture. If some worm larva attached itself to our brains, making us behave in a way that guaranteed death while delivering food, shelter, and sexual orgy to our siblings, we'd probably not think of that parasite as altruistic. But we'd certainly blame it for our misfortune unless we were driven by the desire to help others, a condition psychiatrists call "costly altruism."[6]

That costly altruism story line is infective enough to get *D. dendriticum* held up as a prime example of parasites manipulating their hosts. As is often the case with complex natural phenomena, however, it's far easier to describe behavior-altering infections than it is to demonstrate that they function as expected. *Dicrocoelium dendriticum* is widely distributed throughout Europe, Asia, and North America; its ungulate hosts are regularly slaughtered for food; and when infections are present, sometimes with thousands of worms in the liver, the parasites are readily observed and the livers are discarded. You may not like liver, even with onions, but the agricultural loss amounts to about $2.5 billion globally.[7] The *D. dendriticum* system differs from others involving parasite-induced behavior because the infected ant's jaws lock to the top of grass blades, virtually guaranteeing it will be eaten, rather than just seeming to increase the chances that will happen. So evidently, because sheep eat so much grass that they will eventually get around to grass blades with ants, it's never seemed necessary for scientists to perform experiments demonstrating that sheep preferentially eat ant-laden grass out in the pasture.

But in other cases, parasite-induced behavioral changes seem to just increase the chances that prey containing infective parasite larvae will be captured and thus be consumed more often than uninfected prey, not virtually guarantee it as in the case of *D. dendriticum*. For example, back in the early 1970s, there were published studies on the parasites of ducks in which manipulation of aquatic crustaceans by larval thorny-headed worms played a central role.[8] This research captured quite a bit of attention from biologists, probably because humans are quite familiar with ducks; they show up in cartoons, movies, children's stories, and along the shores of park lakes where people feed them stuff like stale bread that's probably not good for their digestive tracts. Ducks are also extremely attractive game for hunters because they are beautiful, especially in flight; killing them with a shotgun is a fun challenge; they taste great; and if a hunter is successful, they have an opportunity to eat a wild animal and then visit a dentist after biting down on steel shot embedded in breast muscle, breaking a molar.

Ducks have plenty of parasites too, including ones that live in their blood cells, others that live in their blood vessels, and some that live among their feathers. A single duck can have hundreds of tapeworms in its intestine, some of them acquired while it is a duckling on its nesting grounds, others picked up wherever it spends the winter months. In the pond in your local park, ducks drop manure containing worm eggs, some of which hatch into parasites that find their way into snails and then into the skin of anyone wading in that water, causing a serious rash known as swimmer's itch. But what parasitologists interested in behavior love most about ducks is that like the meadowlark in chapter 3, they carry thorny-headed worms, which opened the door for scientific study of the various ways parasites manipulate their hosts.

If you're thinking, "Okay, he's eventually going to tell me that cultural items like smartphones and stupid ideas also manipulate people," you're absolutely correct. But first, here's a truly remarkable observation about thorny-headed worms in the prey that ducks eat.[9] Every duck hunter knows that there are two general kinds of ducks: those that feed mainly on or near the water surface, like mallards, sometimes tipping up so that their rear ends poke up out of the water, and those that dive for food. Regardless of where they get dinner, ducks are not terribly picky about what they eat if it's small enough to go into their mouths. With surface-feeding ducks, their food is mostly aquatic plant material and all the insects, crustaceans, and snails that hang out on it. Diving ducks such as lesser scaup have a similar diet but one that's heavy on invertebrates—snails, aquatic insects, and crustaceans. Surface feeders and divers are opportunistic too, sometimes picking up worms, frogs, and small fish. What else do they eat? The answer is thorny-headed worms, tapeworms, and flukes, parasites that live as immature stages in those insects and crustaceans consumed by ducks. And from parasitologists' research, we learned not only that the crustaceans get manipulated in a way that seems to increase the chances they'll get eaten but also that thorny-headed worms were doing the manipulation in a way that matched the feeding behavior of ducks.

Two different species of thorny-headed worms, both in the genus *Polymorphus*, develop as infective larvae in small freshwater crustaceans named *Gammarus lacustris*, but the two *Polymorphus* species affect their crustacean intermediate hosts differently. When these crustaceans are infected with *P. paradoxus*, which matures in surface-feeding ducks, the crustaceans stay near the surface instead of diving when the water is disturbed, like they would otherwise do with a mallard swishing its bill around; but when infected with *P. marilus*, a parasite of diving ducks, this behavioral change does not occur. Not only do immature thorny-headed worms manipulate their crustacean hosts to increase the chances they will be eaten by a required definitive host, a duck, they do it in a way consistent with the behavior of whatever species of duck their worm matures in.[10] This kind of evolutionary development amazes even parasitologists, who know about so many complex host/parasite relationships that they are not easily amazed.

All this manipulation of hosts by their parasites seems intriguing, consistent with our general impression that parasites are bad, if not outright dangerous, and are thus to be avoided if possible. But scientists studying host manipulation by infectious agents are usually far more interested in the ecological and evolutionary results of such manipulation than on pathology, just one more demonstration that the public perception of science can be quite different from the perception of those doing science. Periodically, however, events happen that eliminate this difference, for example, invasion of some nation by a highly infective virus such as SARS-CoV-2 that forces everyone, including scientists, to be aware of the invader's deadly potential along with all the economic and social impacts that can accompany pandemics. Awareness, of course, is a prelude to behavioral change.

SARS-CoV-2's pathological effects altered host behavior—isolation, mask wearing, and vaccination—even among the uninfected, demonstrating the obvious, that such knowledge is transmitted from host to host, delivered by various media vectors, and, like thorny-headed worms in crustaceans, can function to change behavior. An article by Moises Velasquez-Manoff in the May 25, 2022, issue of the *New York Times Magazine*, for example, explores the impact that misinformation has had on parents' choice to not vaccinate their children against any of the common childhood diseases, in one case sending a doctor printouts of anti-vaccine websites to justify the decision. One Texas pediatrician called this spread of misinformation "the other contagion."[11] And if you tell someone they are behaving in a particular way because they are infected with information, you're thinking and talking like a parasitologist.

Until the early 1980s, the scientific literature contained quite a few demonstrations that infections altered host behavior, and we interpreted those cases as parasite adaptations to increase the probability that a required host would get infected in nature.[12] But someone needed to validate our interpretations with experimental proof; that's how science operates, and not just in the realm of parasitology. Eventually, we need proof that our predictions are correct, and that need applies not only to worms in birds but also to other natural phenomena, climate change being a prime example. And as sometimes happens in science, a graduate student provided that missing proof for parasitic worms. Her name was Janice Moore, and after exploring several other life options, she ended up at the University of New Mexico in the early 1980s, where she decided to study thorny-headed worms in starlings and pill bugs.[13]

To do her planned research, Moore acted on the advice of an ornithologist faculty member at UNM, Dr. David Ligon, who suggested she borrow a technique from Chinese cormorant fishermen to determine exactly what starlings were feeding their nestlings. She chose starlings because they were an abundant invasive species, and thus no federal permit was required to capture them. As Ligon said, "Put up a nest box and you'll get starlings." The technique she borrowed from Chinese fishermen involved gently putting a pipe cleaner around a nestling's neck to prevent it from swallowing something a parent fed it. This technique would allow her to determine exactly what those starling chicks were being fed and ultimately why they were being fed so many pill bugs infected with thorny-headed worms when there were plenty of noninfected pill bugs crawling around in Albuquerque gardens. So Ms. Moore, hoping to become Dr. Moore, set out nest boxes and waited for starlings to arrive and produce babies.[14] Pill bugs—also called roly-polies—were the major players in this effort to understand how these parasites move through and are maintained in an environment. Pill bugs are small crustaceans, mostly less than half an inch long, and terrestrial, meaning they live on the ground instead of in the water, but are more closely related to lobsters than to any of the insects, ticks, and mites in your yard. The most abundant species is named *Armadillidium vulgare*, the *vulgare* part translating into vulgar or common. Why are they called roly-polies? Because they can roll up into a ball when disturbed, sort of like an armadillo, allowing ornery little boys to throw them at one another and shoot them out of BB guns.

Pill bugs get infected with thorny-headed worms when they eat starling feces that contains worm eggs, which subsequently hatch, releasing a larva that burrows through the pill bug intestine and resides inside its body, in the space known as a haemocoel, until its host is eaten by a starling, whereupon it matures, burying its proboscis into the bird's intestinal wall. The worm is named *Plagiorhynchus cylindraceus*, and it's been recognized as a parasite of starlings since the late 1700s. In 1890, Eugene Schieffelin, a member of the American Acclimatization Society, a group formed to bring European plants and animals to the New World, released sixty starlings into Central Park, and in 1891, he released another forty. Rumor has it that Schieffelin thought the United States should have birds mentioned by Shakespeare. Some of those starlings likely had *P. cylindraceus* in their guts.[15] It is generally thought that this entire system—starlings, pill bugs, and *P. cylindraceus*—is imported and that none are native to North America. A century after its purposeful introduction, the American starling population is around two hundred million, distributed throughout the United States, and they're considered invasive pests even though they are beautiful, make great pets if taken from a nest and reared in your kitchen, and have a range of vocalizations that even Mozart is said to have admired.[16]

But the takeaway from this pill bug narrative is that the host/parasite system Moore decided to study is a common one. She just used it in a conceptually powerful way. And that is probably a major lesson to be learned from parasites: even those that do not cause human disease can tell us quite a bit about how infectious agents can move through populations as well as our understanding of life on Earth in general. Parasites and hosts that are accessible, unburdened by regulatory compliance, relatively easy to maintain and manipulate in the lab, and will perform in the lab the same way they do in nature are star players in this scenario. For Janice Moore, the pill bugs, starlings, and thorny-headed worms fulfilled these criteria. And for the rest of the scientific community, that combination was key to legitimizing the study of how parasites manipulate their hosts.[17]

The significance of work that scientists do and the publications that result is not always obvious until we have a historical perspective on it, looking back at the development and evolution of ideas, the changing role that technology plays in research, and the choice of systems to study. For parasitologists, "systems" are combinations of host and parasite species, ideally ones that can be studied both in nature and in the lab and done so in numbers that will satisfy the statistical requirements of publication. With a truly ideal system, a scientist can demonstrate experimentally not only that what seems to happen in nature really does happen but also why it happens. That is exactly what Janice Moore decided to try with pill bugs, starlings, and *P. cylindraceus*, hopefully with the pill bugs performing beautifully, starling parents cooperating, and a parasite manipulating its host in a way that she could demonstrate actually occurred as predicted. Based on those previously published studies involving ducks and aquatic crustaceans, Moore hypothesized that pill bugs infected with thorny-headed worms would behave differently from those not infected. She also hypothesized that the difference in behavior would make infected pill bugs more likely to be eaten by a starling and more likely to be fed to a starling chick than would uninfected pill bugs. Thus, she planned to prove that what had been inferred with the duck research actually happened in nature.

The experiments she devised were elegant and demonstrated conclusively that infected pill bugs behaved differently than uninfected ones, remaining in plain sight rather than seeking shelter and choosing light-colored backgrounds that would make them relatively visible to starlings compared to uninfected ones.[18] The birds responded accordingly, with Moore's use of pipe cleaners being a critical factor in letting her determine in the field that nestlings were being fed a significantly higher number of infected pill bugs than uninfected ones. She was able to watch a starling return to a nest box with food in its bill, then immediately climb up to that nest box, pluck the food from the nestling's mouth before it was able to swallow it, and dissect the food item, which was often a pill bug. With this approach, she answered her question about the effect of parasitism on host behavior and, in doing so, became a pioneer in the study of how parasites manipulate their hosts.

Whatever she accomplished as a scientist, student Moore's example of how to do this research turned her doctoral dissertation into a classic study, setting a standard for what it takes to confirm that your conclusions about parasitism are correct. Moore's published version of this research is classic because she demonstrated that what she observed in laboratory experiments also occurred in nature. Starlings actually picked up infected pill bugs preferentially and fed them to their nestlings. That kind of demonstration is a standard not easily achieved, especially with host/parasite systems, because any one of the participants—eggs, larvae, intermediate hosts, final hosts—can be fickle, tiny, uncooperative, delicate, and wild. These participants can easily require more tender loving care than can even be imagined unless a person has tried to domesticate them, that is, bring them into the lab. Dr. Moore's career flourished after the New Mexico days, especially at Colorado State University where she continued doing research on the ways parasites manipulate their hosts and producing graduate students who also became successful scientists. But the larger takeaway from her time in New Mexico is that demonstration of how and why to design laboratory experiments that produce verified facts from predictions and inferences.

Consideration of the time involved in those starling-pill bug experiments can perhaps help validate this metaphor of cultural events, practices, and innovations as behavior-changing infective agents. To do her experiments showing that infected pill bugs preferred light-colored substrates, making them more visible to starling predators than uninfected ones that sought darker substrates and shelter, Moore had to produce infected pill bugs in the lab. She starved them for four days, fed them carrot pieces covered with worm eggs, then waited three months for worm larvae to become infective before using them in experiments. That worm development time is longer than it took for Donald Trump's claim that the 2020 presidential election had been stolen to manifest itself in the January 6, 2021, attack on the American Capitol building.[19] And only twenty days elapsed between China's closing of the Wuhan market and the first American case of COVID-19;[20] Moore's worms must develop almost five times that long before becoming infective to a starling.

In both these cases—Trump's claim and the COVID-19 invasion—worldwide communication systems functioned as information vectors, transmitting words that were picked up by humans and subsequently altered behavior, even as transportation systems were functioning as physical vectors by moving an infective individual from one nation to another or perpetrators of violence from one part of a nation to another. Trump's assertion recruited groups such as Turning Point USA that helped organize the "Stop the Steal" rally that produced violence, destruction of government property, and deaths on January 6, 2021. In this sense, the recruited supporters functioned exactly like vectors spreading infectious information that produced pathological results. The COVID-19 invasion quickly produced a political response in the form of an anti-vax movement that in turn had a negative public health impact. Anti-vaccination beliefs were not unique to COVID-19, but they were certainly enforced by easy spread of information, again with internet access functioning as a vector.[21]

Thus, both the election claim and the virus revealed behaviors not typically associated with human efforts to avoid danger and even death, but the political response to COVID-19 is especially revealing in this regard. As a consequence, anyone who'd studied the movement of parasites through an ecosystem, regardless of the parasites or hosts, could easily recognize the impact of infectious ideas—vaccines perceived as more dangerous than a disease they prevent and physical barriers to infection, for example masks, perceived as ineffective—with both perceptions carried by words between hosts. Scientists who study disease epidemiology would likely describe the anti-vaccination movement associated with COVID-19 as endemic, meaning always present in a population at some level but often relatively low.[22] That term "endemic" usually also implies a particular geographical area, but that is not the case with this most recent anti-vaccination behavior. Indeed, it, like the virus that spread it, is global.

Sociologists and anthropologists have never had a problem treating cultural items as infective agents capable of manipulating hosts, although they tend to use "innovation" when referring to agents, and they routinely use "transmission" instead of "infection" in their scholarly publications. Chapter 14 explores this concept more thoroughly, although without the mathematics that permeates the scholarly literature on cultural transmission. But that literature also addresses parasitism by ideas that seem to work against the common good, that is, if "good" is defined in terms of health, safety, and relative freedom to pursue various opportunities. Among the many examples of scholars treating ideas as infectious is Glenn Harlan Reynolds's November 20, 2017, column in *USA Today* titled "Social Media Threat: People Learned to Survive Disease, We Can Handle Twitter" and subtitled "We don't know much about the spread of ideas, or what would constitute the equivalent of intellectual indoor plumbing."[23]

Reynolds is a distinguished professor of law at the University of Tennessee; his phrase "intellectual indoor plumbing" is derived from our knowledge of how indoor plumbing and standard of living in general inhibits the spread of disease by erecting a physical barrier between us and microbes or parasites. Reynolds has plenty of company in both the scholarly and popular literature on infective ideas, but his allusion to Twitter in 2017 ties that link to 140 characters, including spaces. He could easily have been referring to a short segment of DNA, coding for forty-six amino acids in some protein, for example part of the SARS-CoV-2 spike protein, the molecule that functions in cell membrane penetration, a requirement before this virus can wreak its internal havoc on some unvaccinated person.

About the time that Janice Moore was feeding thorny-headed worm eggs to pill bugs, Luigi Cavalli-Sforza, an Italian population geneticist from the University of Parma, and Marcus Feldman, a Stanford University biologist, published *Cultural Transmission and Evolution: A Quantitative Approach*, nearly 350 pages of equations addressing the question of "who transmits what to whom?" To answer this question, they drew on data from numerous sources in an attempt to understand "the maintenance of social customs and habits that are spread like infectious diseases." Their data sources included words, and their analysis of word transmission uses language that could easily describe the movement of parasites through a population or between populations. Their analytical techniques—the equations and mathematical models—are similar or identical to those used by epidemiologists.[24]

Shortly after Janice Moore received her PhD, mathematician Douglas Hofstadter published *Metamagical Themas: Questing for the Essence of Mind and Pattern*, in which he discusses the characteristics of what he calls "viral sentences" and gives an example of what words make a phrase infective for humans.[25] That terminology, from 1985, predates the *Oxford English Dictionary*'s definition of "viral" in this sense by four years, but it evidently took another decade for the descriptor to widely infect the English lexicon. We now use the word "viral" more as a descriptor of language, video, and other cultural phenomena than as one of disease, probably because real viruses can have names that seem to sound technical and the diseases end up with their own names, "COVID-19" being a prime example.

Cavalli-Sforza, Feldman, and Hofstadter all published their works before the internet became the avenue for cultural transmission that it is today. Had they been working in the early twenty-first century, their problem would not have been finding examples for analysis but choosing the best examples from the thousands available and actually acquiring data for use in their equations. As this book is being written, two events would provide those three authors with plenty of material to use in the mathematical analysis of information epidemics: the acquisition of social media platform Twitter by billionaire Elon Musk and the defamation lawsuit against Fox News filed by Dominion Voting Systems. In historical terms, both events have a well-defined starting point, a lot of media coverage that can be dated and characterized, and consequences such as termination of Twitter employees and a massive out-of-court settlement that can be considered analogous to pathology.[26]

More recently, Douglas Hofstadter joined with Emmanuel Sander, a cognitive psychologist from the University of Paris (Saint-Denis), to write a book titled *Surfaces and Essences: Analogy as the Fuel and Fire of Thinking*. These authors make an extensively supported claim that analogous words and phrases are highly infective, a good example being that word "viral," which, when used in a sentence, also conveys the ideas of disease, infection, transmission, and transmission mechanisms by inferred analogy—TikTok, for example, spread like a disease-causing virus through the world's population of teenagers. With their analysis of analogy, Hofstadter and Sander provide examples of the mechanism by which words, phrases, and ideas become infectious.[27]

Why are parasitologists especially receptive to the idea that words, phrases, memes, and the items they represent can be infective in the same sense as a virus particle or thorny-headed worms? I believe the answer to that question lies in the complexity of so many parasite life cycles and the diversity of conditions under which animal parasites are transmitted. *Plagiorhynchus cylindraceus*, the worm that carried Janice Moore to her doctorate, has a relatively simple life cycle, requiring only pill bugs and starlings to survive, not a series of hosts like those of some flukes. But the ecological settings in which parasite eggs, pill bugs, and starlings encounter one another, the timing of those encounters, and the fate of the hosts involved can all be quite variable, depending on in which part of the world transmission is happening. In this case, the breadth forced on parasitologists by their discipline opens their minds to metaphors as reality, especially when the behavior of words and worms in populations can be described by the same mathematical equations.

During the week when this last paragraph was written, the CBS program *60 Minutes* had a segment about parents suing the social media platform Instagram for providing images and messages that body-shamed their daughter to the extent that she committed suicide. The extended story on *60 Minutes Overtime* had a lengthy list of online resources for help with similar kinds of crisis. An Instagram program to manage the site for teenagers also began appearing regularly on the social media platform Facebook—an attempt at mental vaccination against image-driven pressure to look like a runway model—and parents can download a guide in thirty-two different languages and eight versions of English to help them help their children deal with a potentially dangerous cultural infection.[28]

When images and text inspire teenagers to harm themselves, those cultural items are acting like pathological infectious agents in the same manner as some parasite that alters behavior, endangering a host. That same week the *60 Minutes* segment was shown,

Greg Abbott, the governor of Texas, banned the use of TikTok on all "government-issued devices throughout all state agencies," according to spectrumlocalnews.com.[29] TikTok was seen as a security risk because of its Chinese ownership. Abbott was clearly acting as if a social media platform was an infection, especially given that computer security software is aimed at preventing infection by electronic viruses, a global problem.

Instagram and TikTok are certainly not the only vectors capable of transmitting infectious ideas that alter human behavior. Indeed, that is the goal of advertising, although the descriptive phrase "transmitting infectious ideas that alter behavior" is never used by advertisers in public or in college classes taught to marketing majors, regardless of whether that intent is expressed in private or in practice. By the time you read this paragraph, there could easily be additional examples similar to the ones mentioned here.

I'm envisioning you complaining to a friend that having read this book, you're now seeing every advertisement as a vector carrying an infective idea parasite, one that someone wants to implant into your brain, rather like those worms in ants, and change your behavior. However, maybe now that you're able to recognize parasitic ideas for what they really are and think of them as potential behavior-altering infections, you'll be able to better evaluate the potential impact that the thousands of communications you receive annually, from hundreds of sources, have not only on your daily life but also on the lives of others. The ducks with their thorny-headed worms and Janice Moore's starlings with theirs remind us that the things we consume are what deliver the infections, although in our case those things are far more varied than tiny crustaceans, and we tend to consume them through our eyes and ears instead of our mouths.

CHAPTER FOURTEEN

GENERAL THEORY OF INFECTIVITY

Words Are Parasites

> If by "infective" you mean whether words can spread like a virus, then yes, words can be infectious in that they can spread from person to person and have an impact on the way people think, feel, and act.
>
> ChatGPT's answer to the question "Are words infective?"[1]

An automobile trip through Iowa in late summer is an agricultural experience best had only once, especially if you are a parasitologist studying the movement of ideas through populations, because the contributions of Iowa farmers to a general theory of infectivity are so visually dominating that it's hard to keep your mind on the highway. Corn stretches as far as you can see, over one rolling hill after another. Depending on how the fields were planted, sometimes you can see down the rows instead of across, the converging lines making a pattern repeated again and again, sometimes mile after mile, as you check your speedometer, avoid dead skunks, and wonder what life is really like for people who live here. When the highway is parallel to rows, you notice immediately that not only are the corn plants all the same height, but their ears are also all at about the same height on the stalks. Iowa corn is bred for the convenience of a machine—the combine—a phenomenon explained by the general theory of infectivity and the diffusion of innovations acting like parasites by taking up residence in people's lives.

Iowa corn farmers have told us much about infectivity, not on purpose but indirectly because they are numerous, about as similar in their behaviors and beliefs as a group of nonrelated humans can be, and all locked into the natural cycles of Planet Earth. In other words, they are a perfect population of hosts. They also communicate with one another, especially in places like the Depot Restaurant and Lounge in Shenandoah, and in recent decades have expanded their information networks via the internet, thus enhancing their ability to spread ideas. Modern agriculture is a highly technical, complex, multifaceted enterprise serviced by both corporations and government at all levels, resulting in a massive database that reveals economic activity, which in turn reveals decision-making on the part of Iowa farmers. So when some scholar decides to study the movement of ideas through a population, those farmers are an ideal group to use as a source of data, for example, to show us how hybrid seed corn was adopted during the mid-twentieth century.

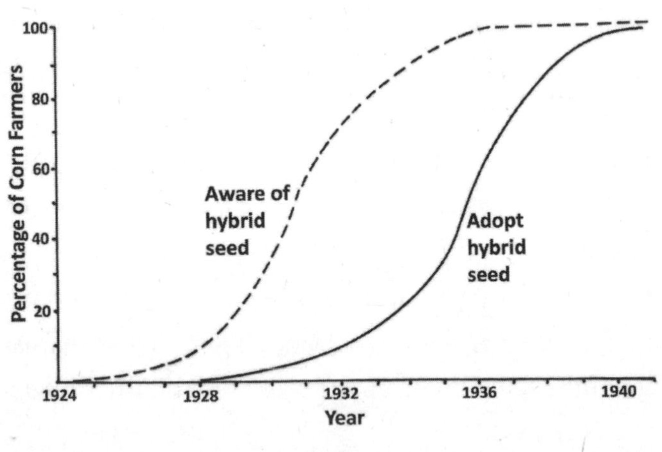

That revelation resulted in what is now widely recognized as a general rule about adoption of innovations, namely that such adoption is best described by an S-shaped curve, as shown at left.

The dotted line shows the percentage of farmers who were aware of hybrid seed corn from the period of 1924 through 1941, and the solid line shows the percentage who actually adopted the innovation. This figure is based on data reported in a 1943 paper written by two faculty members at Iowa State University, Bryce Ryan and Neal Gross, and published in a journal named *Rural Sociology*.[2] To quote from their summary: "Hybrid seed corn has diffused through the Midwest with phenomenal rapidity." The period of 1936 through 1939 was a particularly infective one in which about two-thirds of the farmers either tried hybrid seed on some of their acreage or switched to it altogether. Ryan and Gross may have intended to show us the dynamics of hybrid corn adoption, but the S-shaped curve applies to virtually anything that multiplies until it reaches a limit, whether it be infective technology like smartphones or number of kids with head lice in your child's school.[3]

There has been a literal explosion of research on the diffusion of innovation in the decades since Ryan and Gross did their study and for fairly obvious reasons, the primary one being that data from such adoptions can tell us much about marketing techniques regardless of the product being sold, including political candidates and ideas so outlandish they make us reassess our sense of just how intelligent is our own species. And marketing is, of course, an attempt to start an epidemic in which some infective item moves through a population, altering behavior—writing a check, entering your credit card number, buying a handgun, or voting for someone with no qualifications for public office. Since the mid-1990s, however, once the internet became something close to the global traffic highway it is now, the phrase "diffusion of innovation" has come to apply not only to agricultural products but also to every idea—harebrained or not—every conspiracy theory, every start-up product, every pitch by some wannabe bestselling novelist, and every anything else you can think of that can be spread digitally to anyone with a computer or smartphone anywhere on Earth. The main differences between the spread of these cultural infections and that of parasitic worms are the speed, the mechanisms of transmission, and our reaction to them—you'll rush to buy one while rushing to get rid of the other.

A lot has happened since Ryan and Gross wrote their seminal paper on adoption of hybrid seed corn in the 1930s, and that "a lot" includes not only World War II and the spread of television but also and far more importantly the internet. What may have been an insightful academic exercise for Ryan and Gross is now an intellectual enterprise of staggering proportions and maybe equally staggering importance, ranging from columnists' analysis, sometimes insightful and other times so biased as to defy description, to serious and legitimate scholarly analysis, typically involving mathematical techniques also used in the study of "real" infectious agents such as worms, viruses, or head lice in schoolchildren. We now have an identifiable discipline—memetics—that focuses on problems of information diffusion in populations. It comes as no surprise that those who pursue this discipline have much to tell us about our present immersion in this ocean of words and pictures in various combinations, which few if any of us really know how to navigate, instead using our emotions as guides.

We now ask the question: What do these big Iowa guys, biscuits and sausage gravy on their plates, gun racks in their pickups, and mob–lawyer level smarts about the land in their heads, have to do with a general theory of infectivity? The answer is pretty simple. When one of them says to his neighbor, "You ought to try that hybrid seed," and adds a comment explaining his recommendation, he might as well have sneezed some flu germs, or of late, some SARS-CoV-2 particles. Something in the listener attaches to an infective phrase consisting of four words: "try that hybrid seed." That "something" could easily be the speaker's knowledge and experience, not only recognized by his listeners because they share it but also now providing that phrase—try that hybrid seed—with infective properties it would not otherwise have. The words enter then flow through a neural network. And if the listener asks the speaker to expand on his recommendation, particularly the part that involves money, that's the equivalent, in infectivity terms, of just asking for a couple more sneezes. Words change the behavior of Iowa corn farmers; behavior requires muscle contraction and movement associated directly, if not uniquely, with the uttered, heard, or read phrase and also, probably, in the appearance of the speaker. The first principle of our theory thus becomes:

Words are infective and can alter the physiological condition of a listener.

This first principle is so familiar that you're probably wondering why it is even necessary to consider it part of a general theory, but substitute "worms" for "words" and "pill bugs" for "a listener" and you have Janice Moore's classic research as told in the previous chapter. Adults have all seen this principle in operation but probably didn't think of the word-body interaction as an infection event. All you need to do is step back from your wondering and ask the question: How did I react to the words I've just heard? If those words were spoken by a particular individual in a particular situation, their impact on your emotions would produce different reactions depending on their infectious properties. Three-word sentences are especially illustrative of this point, and here are some familiar examples: "I love you." "Go to hell." "Call me tomorrow." "Send a text." "Meet at Barry's."

If you are over thirty, you've probably heard all these phrases except maybe the last, although you could easily substitute any local watering hole for Barry's Bar and Grill in Lincoln, Nebraska. And your reactions to those sentences varied, depending on who said them, how they were said, when they were said, and the setting in which they were said. The setting and other modifying conditions were given the name "paralanguage" by the anthropologist Peter Farb in his book *Word Play*, an analysis of infectivity but without an epidemiologist's mathematical toolbox.[4] If someone had the right observations, all these whys, whens, hows, and whos could be quantified, although the methods for doing so might raise eyebrows among statisticians more comfortable with their equations being applied to industrial problems than to phrases spreading through a population.

The previous paragraphs suggest rather strongly that words' infective properties can vary significantly over short periods of time and space. In this regard, they are far more variable than familiar parasites such as pinworms, tapeworms, fleas, and lice. In fact, these phrases are more mutable and more rapidly so than the virus made famous by the pandemic of 2020. Furthermore, unlike most if not all evolutionary events, these phrases can mutate back into an original form within seconds. Trust me on this one; try it with "I love you" sometime, although be careful choosing your experimental hosts.

That variability in meaning of a single word or phrase is a fundamental property of language, and thus it becomes a fundamental property of infectious agents in general, given that language is the most infective of all agents that humans encounter on a regular basis. Variation in variants can disappear and then reappear under new environmental conditions, whereas real animals, including all the parasites you've read about so far, vary in evolutionary time instead of seconds or minutes. But words and tapeworms are similar, if not identical, in one respect: conditions of encounter determine whether a potential host becomes infected, and conditions of past encounter determine whether a potential host is resistant or immune.

Again, our phrase "I love you" serves as an excellent illustration of this last point. When parents say it to their children, the meaning is quite different from when a man says it while presenting an engagement ring to the woman sitting across the table with a glass of wine at an upscale restaurant. The children have probably heard that phrase many times and don't react in any particular way, but in the latter case, past experience will determine whether the intended is susceptible or immune to further infection with a diamond ring. Given the culture wars raging across the United States, any of us can imagine a dozen scenarios in which targets of "I love you" are susceptible or immune to infection based on age, gender, jobs, ethnic origin, economic conditions, and past encounters with other humans maybe saying the same thing. Substitute "head lice" or "pinworms" for the "you" in "I love you" and the point is still valid.

The basic epidemiological tool used to analyze movements of infectious agents through populations is called an SEIR model, with the letters referring to fractions of a population: S = susceptible; E = exposed; I = infected; and R = recovered, resistant, or removed. "Removed" can mean "dead."[5] If we know something of the numbers involved, we can estimate the chances that a susceptible individual will become exposed, an exposed individual will become infected, etc. If we apply the correct analytical techniques, we can make predictions about the outcome of an epidemic from some preliminary observations, thus the reference to this tool as a model. It's no surprise that this analytical tool has been applied to the COVID-19 pandemic, although it's also doubtful whether any politicians are studying these scientific papers before making proclamations about the spread of a virus through their constituent population.[6]

Without dragging ourselves into the mathematics involved in actually applying the SEIR model, we can nevertheless readily remember cases in which individuals were susceptible to a suggestion (S); exposed to a piece of information (E); became intellectually infected, that is, believed what they'd read or been told (I); then eventually understood that the information was incorrect (R) or resisted becoming infected with it altogether (also R). In our Iowa corn farmer example, most of the population eventually moved from S through E and eventually to I status. In this case, R can be interpreted as resistant to reverting to S; success of the farmers' hybrid corn adoption meant they'd always use it in the future. A small fraction of them never did adopt hybrid corn and thus were R from the beginning and were never members of S. A sociologist might ask why. You probably know people whose mindset or behaviors would serve as an answer, for example, always voting for candidates of one political party regardless of how qualified those candidates might be for holding any position of responsibility over other humans.

In this example, the words are literal ones, spoken or written in various tracts, including advertising and, in the case of our Iowa corn farmers, agricultural extension division pamphlets. But in this general theory of infectivity, "words" do not need to be spoken or written in any human language; they could easily be nucleotide sequences in DNA, as in a virus particle, a bacterial species, or a tapeworm. In fact, if you could see the whole genetic sequence of some organism, it would likely be a very long string of letters—A, T, C, and G in various combinations—and thus a megaword but still, symbolically if not technically, a word, and one with a meaning. The letters are abbreviations for the nitrogen-containing molecules that comprise DNA. So many parasite species have now been analyzed, at least in part, that scientists regularly match scientific names and these DNA letter sequences, using them almost interchangeably in formal presentations. The "listener" in the case of a DNA word is the cellular machinery that translates that molecular word and puts it into action for all living organisms. So for our theory, the second principle becomes:

Words are any entity that can be identified as, represented by, or linked to a sequence of symbols, signals, and/or spaces in a communication system.

We accept the notion that ideas, thoughts, knowledge, and images as well as sounds, light conditions, and emotions are all real because they can be described, at least partially, or represented by sequences of symbols, space being considered a symbol too. The symbols could be letters in an established language but need not be; they just need to function like letters in some way, particularly by being linked to other entities such as, yes, tangible phenomena (dogs, cats, head lice, tapeworms) or intangible ones such as emotions (love, hate, fear, disgust). Physiological conditions that are or can be altered include all bodily functions—breathing rate, blood pressure, and hormone levels in the blood, to mention a few common ones. The more uncommon ones might include picking up a gun and using it to kill someone or joining an organization that believes relatively light-colored skin makes one superior to other primates in some undefined way regardless of how physically or behaviorally ugly those relatively light-skinned primates might be.

Words, phrases, sounds, ideas, pictures, and the like—the intangible infective words—are typically referred to as memes. Richard Dawkins is generally credited with coining this word "meme," referring to phrases or other intangible objects that move through populations, and introducing it into popular culture in 1976 through his bestselling book *The Selfish Gene*,[7] although anthropologist Peter Farb addressed the concept a year earlier in his book *Word Play: What Happens When People Talk*.[8] Farb also used infectivity as a model, expanding the germ equivalents to include everything from World War II military brides to technology. Humans are mimics; they pick up stuff from other humans, sometimes for rational and easily understood reasons—such as Iowa farmers choosing hybrid seed corn. But sometimes they do so for reasons that defy explanation, or at least seem to until we address that question—What is a human being?—in the most inclusive and imaginative way, not restricting our answers to ones involving sex, money, and power but extending them to ones involving art, music, literature, dance, theater, and every other property or activity we know humans possess, have engaged in over the centuries, and have either copied from or transmitted to other humans, spreading like the flu. We are so variable in our tastes and desires that someone is susceptible, an epidemiological S, to some agent. Most of the time, this susceptibility is harmless and inconsequential as a social phenomenon, but we all know cases in which enough of us are S to particular ideas and suggestions so that collectively we become more of a hazard to civilized society than those head lice on our kids.

We are diverse beyond description, and the parasitic ideas that flow into and out of our minds are equally so, as well as being far more numerous than the 260 different kinds of animal parasites that occupy our bodies around the world. This diversity and flow are just screaming for a general explanatory theory; you're reading my attempt to construct one. The only problem with this attempt is that the arguments are so obvious you may wonder why they need to be explained or codified. The answer to your wondering is that codification produces a system that can be tested with falsifiable hypotheses, although in this case, some of the techniques for such testing may still need to be developed, given the parameters that would have to be measured. Let's take the idea, for example, that drinking wine on Sunday is a sin. I'm guessing that if you tried to spread that idea among a group of acquaintances, you'd figure out in a hurry who was S and who was R. Now, if you tried to make one of your R acquaintances become S, you'd discover very quickly that there are factors either inhibiting or facilitating that infection. In this case, what is the infective agent? It's a string of twenty-six symbols divided into groups by six empty spaces, collectively called "drinking wine on Sunday is a sin." It's a meme. You could probably alter this meme's infective properties by combining it with a picture.

What you've just read is the first simple step in the construction of a general theory of infectivity, namely the lumping of anything that can be represented by a sequence of symbols and spaces into the category of "infective agent." If we were sitting in a bar with a group of parasitologists, you'd likely hear an extension or a further generalization of this practice, for example by referring to anything that could be discussed as a "word." A bacterium is a word; a virus is a word; your kid's pet hamster's DNA sequence makes this rodent a word, although it may also have a name, which means two words are linked to an individual animal and its individual genetic makeup; the music playing on your phone is a word; the original scores of all music are words; some picture that shows up on social media is a word, actually a thousand words if you believe the familiar claim about images as communication devices; gestures you make are words (think raised middle finger); and, of course, you are a word, as is the tapeworm in your gut, if you have one. All you need to do is look in the mirror, reflect on your life to date, the things you've done, your relationships that could be described, your physical attributes, and what others might be saying about you, and suddenly, in addition to your DNA sequence word, you become the equivalent of a novel. Any of the words in that novel could be an infective agent; some of those words could make you rich; others could send you to prison.

There is a reason why people adopt words, phrases, and ways of describing the world around them, and at least in some cases, behaviors are changed as a result of such adoption. If there were not a reason, adoption would not occur, although given the meme involved, discovering that reason ranges from easy and obvious, as with the Iowa corn farmers, to almost impossible or at least difficult and not very satisfactory, as is the case with mob actions stemming from political rhetoric. That is a scientist's rationale at work—seeking a reason even if that reason is as trivial as "someone else said it." The third general principle of our theory thus becomes:

Words enter and leave other words.

This principle can also be expressed as:

Words flow through words.

A flu virus particle makes its way into your lungs and begins to proliferate, making millions more particles; you sneeze, and virus particles leave. Flu virus has flowed through the you word. Viral DNA is a word; your unique DNA sequence is a word. Viral DNA has entered the environmental realm built and maintained by your DNA. You cough; you sneeze; you run a fever; you may vomit or have diarrhea; you take some over-the-counter medicine; you call in sick; you go to bed. The you word tries to kill the flu word, generating some antibodies, and because these antibodies are proteins, sequences of amino acids, they are also words within the context of this discussion. Although I've used you and flu as an example, any of the parasites you've read about so far in this book would have worked equally as well. Thus, we get to the fourth principle of our theory:

Words alter their environments.

The implication of this fourth general principle is that words are alive—they consume energy, occupy space in a mind realm, and reproduce in kind. Any writer would accept that implication, and any biologist will tell you that all living organisms alter the environments in which they occur. In stressful political times, reasons for adoption (words entering their environments) and altered behaviors that result—the analogs of infection and resulting pathology—are or at least can be topics of serious analysis not only by sociologists and psychologists but also by newscasters, media commentators, and editorial writers, all in addition to your next-door neighbor arguing politics over the backyard fence. As an illustration of this point, consider "#MeToo," "Black Lives Matter," and "MAGA." For anyone who watched these memes move through the American population, it was easy to explain their virulence. We understood intuitively why these word objects were so highly infective, an understanding derived from our nation's history of racial and sexual tension. But to generalize a theory of infectivity, we need to also address a wider range of memes, including those that are seemingly harmless yet whose behavior is informative.

As an example of such an infective but politically inconsequential meme with a life cycle as mysterious as that of *Nematobothrium texomensis* in chapter 1, consider the way many longtime residents of the United States use the words "amount" and "number" to describe situations in which multiple items occur. Trust me; the next few sentences hide an infective but intangible word parasite. This parasite will alter the way you interpret your friends' conversations and make you want to correct their use of "amount," and once infected, you will involuntarily react to those conversations. It's quite possible, maybe even probable, that this mental parasite will shape the way you think about your friends forever. And if you correct your friend's usage of these words, that act, a response to infection, may kill the friendship or at least make it ill, symbolically speaking.

You are not alone in your reaction to someone's use of the word "amount"; my Google search on "amount vs. number in conversation" produced 255,000,000 hits in 0.79 seconds. By the time you read this sentence, that number could be much larger than a quarter billion. Trust me again; you are already infected, just from reading the last few sentences, with a meme that will alter your thoughts, your behavior, and probably your opinion of your friends. "Amount vs. number" has infected you, and the first time you talk about this language use to a trusted friend, it will have flowed through you. The extent to which it infects that word friend will depend entirely on whether previous word infections have rendered that word friend susceptible or immune.

The previous paragraph has also altered your awareness level. Unless you are an English teacher or a writer subjected to an editor's red pen, before you read that paragraph, you likely never noticed that some of the people you communicate with use "amount" when they should use "number" or vice versa, and you may have done that yourself. You may have explained tardiness to some social gathering by saying:

"Sorry to be late. I've never seen that amount of cars on the streets at this time of day."

Your friend, however, is cringing inside; she's recently had a conversation with an English teacher who mentioned this word use and is now infected with awareness, looking at you kindly but saying to herself: *number*, you idiot, *number* of cars, not *amount*. You count cars. If you're making cookies, you weigh sugar. Number of cars. Amount of sugar. These last two phrases are going through your friend's mind over and over again, either shaping the way she interprets what you are saying or even blocking her access to the conversation altogether. If that infection produced serious psychopathology, your use of "amount" could easily have altered your friend's evaluation of you as a human being. But your friend's reaction to your words is not too different from what it would be had you sat down in her living room and promptly announced, "I have a really nasty head louse infestation. Hope I don't leave too many nits here on your couch."

From January 20, 2017, to January 19, 2021, the president of the United States used words that altered Americans' evaluation of their fellow human beings, but that alteration has not been as inconsequential as the one resulting from "amount" versus "number."[9] And that president is not alone in his use of language to alter perceptions; we all do it, some of us do it a lot, and depending on how we grew up, we can be mostly unaware of what we are doing. However, after explaining your late arrival to this (pre-COVID or post-vaccination?) social gathering, you say:

"And our son just got admitted to Harvard medical school!" You explain your excitement but decline to mention the lice.

"Amount of cars...amount of cars...amount of cars..." is what your friend is hearing. Ideally, this allergic reaction subsides in time for her to say, "Congratulations!"

The list of entities that move through human populations in a manner analogous to germs is quite long, including almost every item encountered in normal life. All those items have both avenues for dispersal and constraints on their movement. So the real challenge in developing a general theory of infectivity is to come up with assertions that can be tested experimentally or at least observationally across a broad spectrum of infective words, ranging from viruses to tapeworms, fleas, lice, and language intended to start a revolution or dehumanize a starving immigrant trying to escape drug-related violence. Based on our knowledge of animal parasites, here is a list of twenty such assertions, any of which could easily be the basis for doctoral dissertations in linguistics, if they have not already been used in that manner. At least half a dozen have been used repeatedly as the basis for doctoral dissertations in biology. The scientific literature contains numerous examples of these assertions, which are really hypotheses, at work, but finding those examples requires some expertise, time, and energy. So here is an accessible and simplified list:

1. Some words are infective for unexplained reasons.*
2. Some words are infective for unexplainable reasons.*
3. Words evolve, but not necessarily in expected directions.*

4. Words evolve, but not necessarily at expected or predicted rates.*
5. Movement of words through populations can be described mathematically.*
6. Movement of words through populations can be modeled mathematically.*
7. Any identifiable entity can become an infective word.
8. Populations are words, and populations can be made of words.*
9. Words can move through word populations and have no observable effect on any other property of those populations; that is, they can be benign or commensal.*
10. Words can move through word populations and have pathological effects on any other property of those populations; that is, they can be harmful or even deadly.*
11. All means of communication and/or interaction can serve as avenues for word transmission.
12. All words have constraints on their movement through populations.*
13. All words have origins, ontogenies, life cycles, and deaths.*
14. Words vary in their pathological effects on words and word populations, but not always for explained or explainable reasons.*
15. Words can be constructed on purpose but need not be in order to be infective and pathological.
16. Words can be hosts for other words.*
17. Words can be carried by other words without being recognized as parasitic.*
18. Words can immunize hosts against other words.*
19. Words can produce immunodeficiency in hosts.*
20. And finally, words make words (words generate words). This corollary may be the most important of all.*

In the preceding list, the assertions marked with an asterisk (*) are the ones most easily illustrated by the lives of animal parasites, such as tapeworms, fleas, flukes, and protists. Numbers five and six are the ones most commonly found in scientific literature on infectious organisms, and for fairly obvious reasons. Number twelve is the basis for control strategies. Number nine is illustrated by the "amount" versus "number" example above; *Demodex folliculorum* and *D. brevis*, the follicle mites, come pretty close to being real-world examples of that same assertion—a large fraction of the human population is infected, but about the only ones who know it are students taking a parasitology course in which the instructor encourages them to extract such a mite from an eyebrow hair follicle. Figure 41.19 in the textbook *Foundations of Parasitology*, 9th ed., is derived from just such an experience.[10] I doubt if this kind of demonstration is ever written into an animal use protocol (see chapter 8) or would be approved if it were. You can download a free pdf copy of that textbook from the American Society of Parasitologists website.[11]

Assertions ten, eleven, fourteen, and fifteen are easily illustrated by political events of the past century, mainly because of the infection of Earth's human population by technology, ranging from the newspaper and radio to the internet and social media. It could be successfully argued, I believe, within the theory of infectivity, that the internet is the host and humans have infected it, although some would argue, and with considerable reason, that humans are the host species and the internet has infected it. But in terms of technology, smartphones are perhaps the most obvious example of an infectious agent that alters the behavior of its host. That last sentence has been the subject of a massive research enterprise, much of it conducted by various companies' marketing departments.

What remains to be done before we have a general theory of infectivity? The basic ideas need to be tested, of course, but more to the point: someone needs to work at falsifying the predictions of this theory. Such hypothesis testing requires methods that pass scrutiny from those who already study general infectivity, meaning those whose chosen scholarly discipline can be defined as memetics. As is the case with virtually all investigations involving humans, the methods can be a serious challenge given both cultural diversity and individual variation with our population. Platforms generating massive amounts of data have been used to trace the impact of words, for example, using Google to discover responses to pronouncements by Donald Trump on matters such as COVID-19 treatments about which he was totally ignorant.[12]

On another level, it seems perfectly natural to consider our planet to be infected with humans who are in turn infected with many different languages, each providing both avenues for and constraints on the movement of memes through populations, memes that themselves could easily be hosts or vectors for other memes. Parasitologists—or other scientists or readers for that matter—sitting in a bar might look out the window and agree that what they are seeing is an example of nested hyperparasitism, sort of a Russian doll made of words, not all of which are benign. Being parasitologists, however, with their wealth of knowledge about the diversity of infectious agents, the conditions under which they multiply and spread, and the impact they can have on populations, you can be assured that the conversation will eventually turn to the problem of making this planet a better place for all humans, not just a select few. You can also be assured that when this turn occurs, someone will ask: Who could be a vector for this kind of infective language? The unspoken but understood answer will be: Those of us around the table. It's a start.

ACKNOWLEDGMENTS

Life Lessons from a Parasite has turned out to be a relatively complex project, which is not too surprising given that parasitism is manifested in exceptionally diverse ways, at levels ranging from the submicroscopic to the macroscopic and across both geological time and geographical space. For this reason, I've drawn on a large number of published resources and a half century of experience as an academic scientist to support the narrative. I've also drawn on the experiences of my students, ranging from those in Biology 101 to doctoral candidates doing research in my lab, mainly because watching those young people search for parasites, including metaphorical ones, was a remarkable and highly educational experience. Some of the writing is lifted from my previous books, although modified slightly. The central theme of infectivity, including that of ideas, however, was never difficult to find in the stories of scientists asking questions about animals that live in and on other animals. Nor was it difficult to find that central theme in the daily news, especially during elections and the invasion of my nation by a deadly virus.

My wife, Karen, was a patient reader of the original manuscript as well as a much later draft, providing not only a critical eye but also suggestions about how to handle certain topics. Elizabeth Rowson

Zahorjan, a freelance editor, did excellent work on the introduction and chapter 1. Two student readers, Sarah Hawkinson and Kaylen Michaelis, provided commentary on the early version; both students were taking an English course titled Literature and Environment at the University of Nebraska's Cedar Point Biological Station in western Nebraska. I was invited to visit that class and asked if anyone would like to read this work in progress; both expressed interest so I sent them each an electronic file.

Following her initial reading, I was able to hire Kaylen Michaelis for editorial and research help, and she did serious editorial work on the book over a two-year period, working at the professional level. Even though she was an undergraduate student at the University of Nebraska–Lincoln, she was in the Honors Program, a four-year Regents Scholar, and worked also in the English Department's Writing Center. She had a special color-coded approach to her work, whether it was on paper manuscript or digital files, and her comments ranged from punctuation and grammar corrections to insightful observations on what I was trying to accomplish and how best to accomplish it. I've worked with editors from several major publishers, and Kaylen was clearly in their league. When she graduated, Kaylen recommended a replacement, Angelina Pattavina, a college sophomore at the time but also a superb editor who worked on another project and also read the *Life Lessons* page proof.

I also express deep appreciation for other friends and colleagues who read the manuscript and provided both comments and endorsements. These individuals include Dr. Fred Ohles, retired president of Nebraska Wesleyan University; Dr. Chuck Blend, scientist/educator at the Museum of Science & History in Corpus Christi, Texas, Natural History Museum; and Dr. Richard Clopton, professor of biology at Peru State College and editor of the *Journal of Parasitology*. Dr. Clopton gave me permission to use excerpts from papers published in the *Journal of Parasitology*. He also read later versions of chapters in which he is discussed at length and had no objections to the narrative.

Dr. Janice Moore, retired professor from Colorado State University, submitted to a Zoom interview for chapter 13, read a draft, and suggested changes that I incorporated. I have known her for many years through scientific meetings and have always admired her creativity and intellectual courage. That work on parasite manipulation of host behavior is a classic. Dr. Janine Caira, University of Connecticut, read chapter 12, in which she is mentioned, and approved of the way I described her research. Dr. Matthew Bolek, Oklahoma State University, also read chapter 11, based on his research, and approved of it.

Leslie Meredith, my agent at Dystel, Goderich, and Bourret, was of major help and encouragement, editing the proposal as well as some of the chapters and handling the submissions in a timely and objective way. The copy editors Emily Proano and Sabrina Baskey did a superb job, saving me from some embarrassing mistakes. Finally, my editor at Sourcebooks, Jenna Jankowski, handled this project in an extremely professional manner, being patient and insightful with my drafts, and like the best editors, she ended up making it a far better book than it would have been without her work. She was especially adept at seeing ways for the narrative to connect with readers other than my parasitologist friends, thus making the ideas far more accessible than they might have been otherwise. In essence, she turned the book I wanted to write into a book that needed to be written.

A CONVERSATION WITH THE AUTHOR

Let's start with the obvious: Was there a particular moment that inspired you to write this book?

At lunch with a couple of retired colleagues a few years ago, we had a conversation about what we'd learned from our lifetime research experiences, and as a result of that lunch talk, I started thinking that this book or something similar to it needed to be written. The other two scientists were not parasitologists, but we all ended up talking about how our research taught us more about science in general as a human activity than about parasites, physiology, or molecular biology. The political campaign of 2016 was also a major stimulus to finish this book, mainly because the internet seemed to have completely changed the way ideas moved through populations. I found myself thinking almost continuously that I was seeing an electronic vector spreading infectious words that in turn altered behavior—all clearly analogous to what happens with many animal parasites in nature. That analogy is now embedded throughout *Life Lessons from a Parasite*.

Parasites are a maligned group. What would you say is the public's biggest misconception about them?

This biggest misconception is that all parasites are dangerous and pathological. Well, they do take some of their hosts' resources, of course, and some of them are quite pathological, but the vast majority of them are relatively harmless, at least compared to predators and environmental destruction. We need to remember that parasitism is the most common way of life among animals. If you take a field trip or even watch birds in your backyard, most of the animals you see will have at least one kind of parasite, and the birds in particular are likely to have several kinds of other animals that live in and on them. Even insects have parasites. There are relatively few parasites that cause major human diseases, and those species are indeed important economic factors, especially in the tropics, but the tapeworms in your backyard birds don't have much of an impact on how successful those birds are at rearing their young, and the flea-scratching squirrels are still quite active and rearing young that eventually acquire their own fleas, probably from their parents, just like human children acquire habits and values from their own parents.

What, if anything, sets parasitology apart from other biological disciplines?

I believe that the diversity of parasitic relationships and the fact that you can't really learn much about a parasite without also studying its hosts requires a broad approach that is not necessarily present in other disciplines. The complex life cycles of many parasites, especially when combined with ecological factors influencing transmission, also contribute to this diversity. The number of scientific names that a student might encounter in a parasitology course, for example, and the diversity of organisms these names refer to far exceed the number of names and diversity of organisms encountered in an ornithology course, and each of those names carries all the historical, taxonomic, ecological, and evolutionary baggage of any scientific name. That understanding of diversity—how it arises and what it means—gets transferred to areas of our lives far beyond the research lab.

What would you say to a squeamish reader to make them care about parasites? How do you get folks interested in the subject?

I believe that by equating infectious ideas to infectious worms and focusing on transmission, we can get to the truly interesting aspects of parasitism very quickly. And equating certain infectious ideas to behavior-altering parasitic infections moves the conversation beyond squeamishness to curiosity and then sometimes to those interesting subjects like politics and religion, in which infective ideas are central.

Can you share some of your favorite influences or references that shaped your work and thinking?

Aside from important mentors, three works from the 1970s and '80s shaped my whole approach to teaching and how I view the world: Peter Farb's *Word Play*, Douglas Hofstadter's *Metamagical Themas*, and Luigi Cavalli-Sforza and Marcus Feldman's *Cultural Transmission and Evolution*. All three are mentioned in *Life Lessons*, in the later chapters. These three works established in my mind the concept of ideas and cultural items as infective agents. Farb's *Word Play* especially was like a slap on the head; I remember thinking, wow, words, language, and the ideas they convey function just like disease agents moving through populations. After encountering these works, everything I did as a teacher was designed to increase the infectivity of biological subjects and alter the way my students viewed their environments. My main grad school mentors, J. Teague Self and George M. Sutton, were also really role models on how to be a successful college prof, especially by trying to produce students who were in turn successful.

NOTES

There are numerous entries below for *Foundations of Parasitology*, 9th edition, by Larry S. Roberts, John Janovy Jr., and Steven A. Nadler. The rights were returned to the authors by McGraw-Hill, and that textbook is now available as a free pdf download from the American Society of Parasitologists' website: https://www.amsocparasit.org/.

Introduction

1 "Head Lice Treatment," Centers for Disease Control and Prevention, U.S. Department of Health and Human Services, last updated October 15, 2019, https://www.cdc.gov/parasites/lice/head/treatment.html.

2 "Head Lice," Mayo Clinic, last updated April 30, 2022, https://www.mayoclinic.org/diseases-conditions/head-lice/diagnosis-treatment/drc-20356186.

3 Robin A. Weiss, "Apes, Lice and Prehistory," *Journal of Biology* 8, no. 2 (2009): 20, https://doi.org/10.1186%2Fjbiol114.

4 C. Lawrence Cooper and John L. Crites, "A Checklist of the Helminth Parasites of the Robin, Turdus migratorius Ridgway," *American Midland Naturalist* 95, no. 1 (January 1976): 194–96, https://doi.org/10.2307/2424246.

5 Larry S. Roberts, John Janovy Jr., and Steven A. Nadler, *Foundations of Parasitology*, 9th ed. (New York: McGraw-Hill, 2012), 1–8.

6 A. E. R. Westman, "Poem—A Chemist Looks at Parasitology," *Journal of Parasitology* 54 (1972): 698.

Chapter One: Why Parasites?

I made a major effort to find information on the life of Clarence E. Davis Jr. after he left University of Oklahoma and returned to Southern University, writing both emails and letters to anyone I believed might have such information. These sources included names that were found in the field notes. Some of these individuals are deceased, and others were either unable or unwilling to answer my questions; several did not respond. I owe special thanks to Dr. Scott Gardner, curator of parasitology at the Harold W. Manter Laboratory of Parasitology, University of Nebraska State Museum, for locating the Self field notes and letting me scan them. I also thank the collections manager in that laboratory, Dr. Gabor Racz, for his help with access to specimens and photography.

1 Gerald W. Esch, *Parasites, People, and Places: Essays on Field Parasitology* (New York: Cambridge University Press, 2004), 148.

2 Allen McIntosh and J. Teague Self, "*Nematobothrium texomensis* n. sp. from a Freshwater Fish, *Ictiobus bubalus*," *Journal of Parasitology* 41 (Sec. 2, Abstract #83, 1955): 36–37; J. Teague Self, Lewis E. Peters, and Clarence E. Davis, "The Egg, Miracidium, and Adult of *Nematobothrium texomensis* (Trematoda: Digenea)," *Journal of Parasitology* 49, no. 5 (1963): 731–36, https://doi.org/10.2307/3275914 J. Teague Self, Lewis E. Peters, and Clarence E. Davis, "The Biology of *Nematobothrium texomensis* McIntosh and Self, 1955 (Didymozoidae) in the Buffalo Fishes of Lake Texoma," *Journal of Parasitology* 47 (Sec. 2, Abstract #92, 1961): 42–43.

3 "Lake Texoma," Lake Texoma Association, last updated June 20, 2019, https://www.laketexomaonline.com/.

4 McIntosh and Self, "*Nematobothrium texomensis*," 36–37; Self, Peters, and Davis, "Egg, Miracidium, and Adult," 731–36; Self, Peters, and Davis, "Biology of *Nematobothrium texomensis*," 42–43.

5 Roberts, Janovy, and Nadler, *Foundations of Parasitology*, 201–348.

6 Self, Peters, and Davis, "Biology of *Nematobothrium texomensis*."

7 J. Teague Self and Students, "Field Notebook-Parasitology" (unpublished book manuscript, started March 1960), Harold W. Manter Laboratory Collections, University of Nebraska State Museum.

8 McIntosh and Self, "*Nematobothrium texomensis*."

9 Self and Students, "Field Notebook-Parasitology."

10 Roberts, Janovy, and Nadler, *Foundations of Parasitology*, 201–348.

11 Roberts, Janovy, and Nadler, *Foundations of Parasitology*, 201–348.

12 Self and Students, "Field Notebook-Parasitology."

13 Clarisse Louvard et al., "First Elucidation of a Didymozoid Life Cycle: *Saccularina magnacetabula* n. gen. n. sp. Infecting an Arcid Bivalve," *International Journal for Parasitology* 52, no. 7 (June 2022): 407–25, https://doi.org/10.1016/j.ijpara.2021.11.012.

Chapter Two: Cheyenne Bottoms

This chapter is based largely on my experience as a research assistant on the Bird/Virus/Parasite project funded by the National Institutes of Allergy and Infectious Diseases in the mid-1960s and conducted by Dr. J. T. Self at the University of Oklahoma, Dr. Vernon Scott at the OU Medical Center, and Dr. David Parmelee from Emporia State University. They referred to this project as the BVP, and it also funded my doctoral dissertation research on avian malaria. I spent three years going to the Cheyenne Bottoms, a sixty-four-square mile wetland derived from a geological sinkhole a few miles east of Great Bend, Kansas. Special thanks to Courtney Kennedy, Student Success Team at the University of Oklahoma Libraries, for supplying a copy of Dan Harlow's MS thesis.

1 Dan Rae Harlow, "The Helminth Fauna of the Pectoral Sandpiper (*Erolia melanotos*) with Special Reference to the Effects of Migration" (master's thesis, University of Oklahoma, 1962), 1.

2 John L. Zimmerman, *Cheyenne Bottoms: Wetland in Jeopardy* (Lawrence: University Press of Kansas, 1990), 13–24.

3 Wikipedia, s.v. "Cheyenne Bottoms," last updated May 20, 2022, https://en.wikipedia.org/wiki/Cheyenne_Bottoms.

4 Klaudia Adamczewska-Chmiel et al., "Smartphones, the Epidemic of the 21st Century: A Possible Source of Addictions and Neuropsychiatric Consequences," *International Journal of Environmental Research* 19, no. 9 (2022): 5152, https://doi.org/10.3390/ijerph19095152.

5 Harlow, "Helminth Fauna of the Pectoral Sandpiper," 1.

6 "Eastern Equine Encephalitis Virus," Centers for Disease Control and Prevention, U.S. Department of Health and Human Services, last updated October 31, 2022, https://www.cdc.gov/easternequineencephalitis/symptoms-diagnosis-treatment/index.html; "Western Equine Encephalitis Virus Fact Sheet," Minnesota Department of Health, last updated March 1, 2018, https://www.health.state.mn.us/diseases/weencephalitis/wee.html.

7 Mary Jo Hellmer, Roy W. McGuire, and L. Vernon Scott, "A Preliminary Report on Isolations of Arboviruses from Mosquitoes and Migratory Birds," *Proceedings of the Oklahoma Academy of Sciences* 45 (1965): 227–28, https://ojs.library.okstate.edu/osu/index.php/OAS/article/view/4393.

8 Hellmer, McGuire, and Scott, "Preliminary Report," 227–228.

9 "What to Do about Starlings," Humane Society of the United States,

last updated July 1, 2022, https://www.humanesociety.org/resources/what-do-about-starlings.

10 John Janovy Jr., "Epidemiology of Malaria in Certain Birds of the Cheyenne Bottoms, Barton County, Kansas" (PhD dissertation, University of Oklahoma, 1965).

11 "Baird's Sandpiper," Cornell Lab of Ornithology, All About Birds, accessed August 29, 2023, https://www.allaboutbirds.org/guide/Bairds_Sandpiper/overview.

12 "Pectoral Sandpiper," Cornell Lab of Ornithology, All About Birds, accessed August 30, 2023, https://www.allaboutbirds.org/guide/Pectoral_Sandpiper.

13 James Dennison and Andrew Geddes, "Thinking Globally about Attitudes to Immigration: Concerns about Social Conflict, Economic Competition and Cultural Threat," *Political Quarterly* 92, no. 3 (2021): 541–51, https://doi.org/10.1111/1467-923X.13013.

Chapter Three: Meadowlark

My position as research assistant to the BVP resulted in the handling of several hundred birds of six different species. I dissected them all, went through their intestines and lungs looking for parasites, and studied their blood smears. Although ectoparasites were not included in the study, I did collect many of them and made prepared slides from the specimens. The paid collector was a man named Homer A. "Steve" Stephens who, when I knew him, lived with a pet prairie dog in a pickup camper of his own construction and traveled the state of Kansas studying woody vegetation for a book he was writing. Gabor Racz, collections manager at the Harold W. Manter Laboratory of Parasitology, University of Nebraska State Museum, took the photomicrographs mentioned in this chapter. There are several websites that provide the epigraph quote from Edmond Waller.

1 "Poetical Works of Edmund Waller and Sir John Denham. With Memoir and Dissertation, by the Rev. George Gilfillan," Project Gutenberg, accessed January 12, 2024, https://www.gutenberg.org/ebooks/12322.

2 "Differences between Eastern Meadowlark, *Sturnella magna*, and Western Meadowlark, *S. neglecta*," Cornell University Museum of Vertebrates, last updated November 2, 2021, https://www.birds.cornell .edu/crows/mlarkdiff.htm.

3 "Plant 3—Lincoln, Nebraska," Elgin.Watch, accessed August 29, 2023, https://www.elgin.watch/enwco/plants/plant-lincoln/.

4 Donald V. Moore, "Morphology, Life History, and Development of the Acanthocephalan *Mediorhynchus grandis* Van Cleave, 1916," *Journal of Parasitology* 48, no. 1 (February 1962): 76–86, https://doi .org/10.2307/3275416.

5 Malte Sielaff et al., "Phylogeny of Syndermata (syn. Rotifera): Mitochondrial Gene Order Verifies Epizoic Seisonidea as Sister to Endoparasitic Acanthocephala within Monophyletic Hemirotifera," *Molecular Phylogenetics and Evolution* 96 (March 2016): 79–92, https:// doi.org/10.1016/j.ympev.2015.11.017.

Chapter Four: Favorite Maggots

During the years from 1967 through 1982, our research on parasitic flagellates was supported by the Department of the Army, the World Health Organization, and the National Science Foundation. That support allowed us to do some truly creative work with parasite species originally isolated and cultured not only from humans and lizards but also from insects. The diverse origins of these related parasites allowed us to do significant comparative work, although the students were the ones who actually did most of the late-night labor. The bracelet mentioned at the end of this chapter was made by Jackie Lusher, a Lincoln, NE, jewelry maker, now deceased.

"The Hearse Song" is traditional with no generally accepted origin, although the "worms crawl in, the worms crawl out" language can evidently be traced to a 1796 novel titled *The Monk*, written by Matthew Lewis. The song has various versions and was sung by both British and American soldiers during World War I.

1 Wikipedia, s.v. "The Hearse Song," last updated July 5, 2023, https://en.wikipedia.org/wiki/The_Hearse_Song.

2 Roberts, Janovy, and Nadler, *Foundations of Parasitology*, 592–98.

3 Rui Chen et al., "*Megaselia scalaris*, Scuttle Fly (Diptera: Phoridae)," LSU Agricultural Center, accessed August 30, 2023, https://www.lsuagcenter.com/~/media/system/1/e/3/f/1e3fe4d44d75da57e962f042ddad3334/p3850_bugbizscuttlefly_1122pdf.pdf.

4 Roberts, Janovy, and Nadler, *Foundations of Parasitology*, 61–86.

5 George Poinar Jr. and Roberta Poinar, "*Paleoleishmania proterus* n.gen., n. sp., (Trypanosomatidae: Kinetoplastida) from Cretaceous Burmese Amber," *Protist* 155, no. 3 (September 2004): 305–10, https://doi.org/10.1078/1434461041844259.

6 Norman R. Dollahon and John Janovy Jr., "Experimental Infection of New World Lizards with Old World Lizard *Leishmania* Species," *Experimental Parasitology* 36, no. 2 (October 1974): 253–60, https://doi.org/10.1016/0014-4894(74)90064-2.

7 Pierre Marc Daggett, Norman R. Dollahon, and John Janovy Jr., "*Herpetomonas megaseliae* sp. n. (Protozoa: Trypanosomatidae) from *Megaselia scalaris* (Loew, 1866) Schmitz, 1929 (Diptera: Phoridae)," *Journal of Parasitology* 58, no. 5 (October 1972): 946–49, https://doi.org/10.2307/3286591.

8 Daggett, Dollahon, and Janovy, "*Herpetomonas megaseliae*," 946–49.

9 Pierre Marc Daggett, Joan E. Decker, and John Janovy Jr.,

"Some Physiological Alterations Accompanying Infectivity to Mammals by Four Genera of Trypanosomatidae," *Comparative Biochemistry and Physiology* 59, no. 4 (1978): 363–66, https://doi .org/10.1016/0300-9629(78)90178-0.

10 Norman R. Dollahon and John Janovy Jr., "Insect Flagellates from Feces and Gut Contents of Four Genera of Lizards," *Journal of Parasitology* 57, no. 5 (October 1971): 1130–32, https://doi.org/10.2307/3277877.

Chapter Five: The Plains Killifish

The University of Nebraska's Cedar Point Biological Station (CPBS), eight miles north of Ogallala, in Keith County, is the setting for this chapter. The generations of students who studied parasites in small fishes were amazing in their efforts and insights. My students and I spent nearly thirty years studying the parasite fauna of small fishes, including *Fundulus zebrinus*, in western Nebraska. That work produced new species descriptions, both theoretical and ecological papers, and a vision of the natural world that centered on a braided prairie river with highly variable streamflow as both a fact and a metaphor. I greatly appreciate the opportunity to be a part of the CPBS program.

Of the two students mentioned at the beginning of this chapter, both ended up as physicians; Mike McCarty is deceased; Rick Goble is retired.

1 John Janovy Jr., "Field Note: The Inception of the CPBS and Our First Fish," American Society of Parasitologists, last updated December 12, 2018, https://www.amsocparasit.org/post/field-note-the -inception-of-the-cpbs-and-our-first-fish.

2 "Platte River Near Grand Island, Nebr.-06770500," United States Geological Survey, last updated August 29, 2023, https://waterdata

.usgs.gov/monitoring-location/06770500/#parameterCode=00065&period=P7D.

3 "South Platte River Snow Report," SnoFlo, January 8, 2023, https://snoflo.org/report/snow/colorado/south-platte-river/.

4 John Janovy Jr. and Eugene Lee Hardin, "Population Dynamics of Parasites in *Fundulus zebrinus* in the Platte River of Nebraska," *Journal of Parasitology* 73, no. 4 (August 1987): 689–96, https://doi.org/10.2307/3282396; John Janovy Jr. and Eugene Lee Hardin, "Diversity of the Parasite Assemblage of *Fundulus zebrinus* in the Platte River of Nebraska," *Journal of Parasitology* 74, no. 2 (April 1988): 207–13, https://doi.org/10.2307/3282446; John Janovy Jr., Scott D. Snyder, and Richard E. Clopton, "Evolutionary Constraints on Population Structure: The Parasites of *Fundulus zebrinus* (Pisces: Cyprinodontidae) in the South Platte River of Nebraska," *Journal of Parasitology* 83, no. 4 (August 1997): 584–92, https://doi.org/10.2307/3284228.

5 Permission to use this excerpt granted by Richard E. Clopton, editor of *Journal of Parasitology*, in an email dated May 21, 2021.

6 Deborah D'Souza, "TikTok: What It Is, How It Works, and Why It's Popular," Investopedia, last updated August 14, 2023, https://www.investopedia.com/what-is-tiktok-4588933.

7 David Kiley, "A New NAFTA Agreement That Includes Canada Is Likely This Week," *Forbes*, August 29, 2018, https://www.forbes.com/sites/davidkiley5/2018/08/29/a-new-nafta-agreement-that-includes-canada-is-likely-this-week/?sh=739fb83c3327; David Leggett, "Auto Industry Ready to Witness Job Losses and Major Skill Transitioning," JustAuto, October 28, 2021, https://www.just-auto.com/features/auto-industry-ready-to-witness-job-losses-and-major-skill-transitioning/.

8 Janovy and Hardin, "Population Dynamics," 689–96.

9 "What Are El Niño and La Niña?," National Ocean Service, National Oceanic and Atmospheric Administration, last updated August 24, 2023, https://oceanservice.noaa.gov/facts/ninonina.html.

10 Mirna Alsharif and Ray Sanchez, "Bodies of Covid-19 Victims Are Still Stored in Refrigerated Trucks in NYC," CNN, last updated May 7, 2021, https://www.cnn.com/2021/05/07/us/new-york-coronavirus-victims-refrigerated-trucks/index.html; Alessandro Santini et al., "COVID-19: Dealing with Ventilator Shortage," *Current Opinion in Critical Care* 28, no. 6 (December 2022): 65259, https://doi.org/10.1097/MCC.0000000000001000; Jill Neimark, "The Extra Deaths," OpenMind, last updated April 1, 2022, https://www.openmindmag.org/articles/the-extra-deaths.

11 John Janovy Jr., Timothy R. Ruhnke, and Terry A. Wheeler, "*Salsuginus thalkeni* n. sp. (Monogenea: Ancyrocephalidae) from *Fundulus zebrinus* in the South Platte River of Nebraska," *Journal of Parasitology* 75, no. 3 (June 1989): 344–47, https://doi.org/10.2307/3282584.

12 Janovy, Snyder, and Clopton, "Evolutionary Constraints," 584–92.

13 Rebecca Lindsey, "Climate Change: Global Sea Level," National Oceanic and Atmospheric Administration, last updated April 19, 2022, https://www.climate.gov/news-features/understanding-climate/climate-change-global-sea-level; Lawrence S. Kalkstein and Karen E. Smoyer, "The Impact of Climate Change on Human Health: Some International Implications," *Experientia* 49, no. 11 (November 1993): 969–79, https://doi.org/10.1007/bf02125644; Anthony J. McMichael, Rosalie E. Woodruff, and Simon Hales, "Climate Change and Human Health: Present and Future Risks," *Lancet* 367, no. 9513 (March 2006): 859–69, https://doi.org/10.1016/S1040-6736(06)68079-3.

14 Ari Drennen and Sally Hardin, "Climate Deniers in the 117th

Congress," *Center for American Progress*, March 30, 2021, https://www.americanprogress.org/article/climate-deniers-117th-congress/.

15 Karen Armstrong, *The Battle for God: Fundamentalism in Judaism, Christianity and Islam* (New York: Random House, 2000), 167–73.

16 "Migration: An In-Depth Collection of Global Reporting on Refugees, Asylum Seekers, Migrants, and Internally Displaced People," New Humanitarian, accessed August 29, 2023, https://www.thenewhumanitarian.org/migration.

Chapter Six: Rocky Mountain Toad

The student in this chapter is Eugene Lee Hardin, who enrolled in my Field Parasitology course at the Cedar Point Biological Station (CPBS) and later did master's thesis research in my laboratory. He is now a physician. Matthew Bolek, the subject of chapter 11, also worked on the life cycle of the tapeworm discussed in this chapter, as did his students in another course taught at CPBS, mostly to no avail. The American Society of Parasitologists' web page has a story about Bolek and his students' attempts to solve the life cycle; link is given below. The details of this tapeworm's life remain a mystery.

There are various wordings of this quote from Bion of Borysthenes (c. 325–250 BC), a freed slave who later became a philosopher and writer. This wording of the quote seemed to be more appropriate for the chapter than did some others.

1 Wikiquote, s.v. "Bion of Borysthenes," last updated May 3, 2021, https://en.wikiquote.org/wiki/Bion_of_Borysthenes.

2 Hans Zinsser, *Rats, Lice, and History: Being a Study in Biography, Which, after Twelve Preliminary Chapters Indispensable for the Preparation of the Lay Reader, Deals with the Life History of Typhus Fever* (Boston: Atlantic Monthly Press, 1963), 133.

3 Lloyd B. Dickey, "A New Amphibian Cestode," *Journal of Parasitology* 7, no. 3 (March 1921): 129–36, https://doi.org/10.2307/3270780.

4 Wikipedia, s.v. "Rocky Mountain toad," last updated February 26, 2021, https://en.wikipedia.org/wiki/Rocky_Mountain_toad.

5 Stephen R. Goldberg, Charles R. Bursey, and Irma Ramos, "The Component Parasite Community of Three Sympatric Toad Species, *Bufo cognatus*, *Bufo debilis* (Bufonidae), and *Spea multiplicata* (Pelobatidae) from New Mexico," *Journal of the Helminthological Society of Washington* 62, no. 1 (1995): 57–61; Stephen R. Goldberg and Sonia Hernandez, "Helminths of the Western Toad, *Bufo boreas* (Bufonidae) from Southern California," *Bulletin of the Southern California Academy of Sciences* 98, no. 1 (1999): 39–44.

6 Matthew Bolek, "Field Note: Attempting to Solve the Life Cycle of Distoichometra bufonis," *American Society of Parasitologists*, November 15, 2018, https://www.amsocparasit.org/post/field-note-attempting-to-solve-the-life-cycle-of-distoichometra-bufonis.

7 Wikipedia, s.v. "Rocky Mountain toad."

8 Eugene Lee Hardin, "The Population Biology of *Distoichometra bufonis* Dickey, 1921 (Cyclophyllidea: Nematotaeniidae) in *Bufo woodhousii* Girard, 1854 (Amphibia: Bufonidae)" (master's thesis, University of Nebraska–Lincoln, 1987); Eugene Lee Hardin and John Janovy Jr., "Population Dynamics of *Distoichometra bufonis* (Cestoda: Nematotaeniidae) in *Bufo woodhousii*," *Journal of Parasitology* 74, no. 3 (June 1988): 360–65, https://doi.org/10.2307/3282038.

9 "U.S. Immigration Trends," Migration Policy Institute, last updated May 10, 2023, https://www.migrationpolicy.org/programs/data-hub/us-immigration-trends; "Climate Change and Invasive Species," National Invasive Species Awareness Week, last updated March 8, 2023, https://www.nisaw.org/climatechange/.

Chapter Seven: Death of a Beetle

This essay was originally written for and delivered to students in a large introductory biology class at the University of Nebraska–Lincoln. There were about 260 students in that lecture section. The mealworm narrative is based on decades of teaching and research using parasites of insects, especially the beetle *Tenebrio molitor*. The definitive work on these parasites was done by Dr. Richard Clopton, now a professor at Peru State College but at the time a doctoral student in my lab. He subsequently had a career as a highly successful biologist, publishing extensively and being editor of both *Comparative Parasitology* and *Journal of Parasitology*. Although this essay was written many years ago, internet sources below address the same causes of death in terms of current statistics.

1 Lewis Thomas, *Lives of a Cell: Notes of a Biology Watcher* (New York: Viking, 1974), 96–99.

2 Paul Fussell, *Wartime: Understanding and Behavior in the Second World War* (New York: Oxford University Press, 1989), 164–80; Edward O. Wilson, *On Human Nature* (Cambridge, MA: Harvard University Press, 1978), 113–14.

3 John Gramlich, “What the Data Says about Gun Deaths in the U.S.,” Pew Research Center, last updated April 26, 2023, https://www.pewresearch.org/fact-tank/2022/02/03/what-the-data-says-about-gun-deaths-in-the-u-s/.

4 “Fatality Facts 2020 State by State,” Insurance Institute for Highway Safety, last updated May 1, 2023, https://www.iihs.org/topics/fatality-statistics/detail/state-by-state.

5 Gerald Urquhart et al., “Tropical Deforestation,” National Aeronautics and Space Administration Earth Observatory, last updated March 30, 2007, https://earthobservatory

.nasa.gov/ContentFeature/Deforestation/tropical_deforestation_2001 .pdf.

6 Jason Noble, "Iowa 3rd in Chances of Hitting Deer with Vehicle," *Des Moines Register*, October 1, 2013, https://www .desmoinesregister.com/story/news/1/01/01/iowa-3rd-in-chances -of-hitting-deer-with-vehicle/2906593/.

7 "Nebraska Speeding and Car Accidents," Robert Pahlke Law Group, last updated September 26, 2019, https://www.pahlkelawgroup .com/personal-injury/car-accidents/speeding/.

8 Richard E. Clopton, Tamara J. Percival, and John Janovy Jr., "*Gregarina niphandrodes* n. sp. (Apicomplexa: Eugregarinorida) from Adult *Tenebrio molitor* (L.) with Oocyst Descriptions of Other Gregarine Parasites of the Yellow Mealworm," *Journal of Protozoology* 38, no. 5 (September 1991): 472–79, https://doi.org/10.1111/j.1550-7408.1991 .tb04819.x.

Chapter Eight: An Anti-Desensitization Plan

This chapter is a product of events that occurred at the University of Nebraska–Lincoln during the fall of 2017, when someone who had evidently been at the Cedar Point Biological Station (CPBS) in Keith County, Nebraska, communicated with the Institutional Animal Care and Use Committee, indicating that certain provisions of an approved protocol had been violated. These violations were alleged to have occurred during Field Parasitology, one of the CPBS courses, taught by Dr. Scott Gardner, the curator of parasitology at the University of Nebraska State Museum. The resulting events and communications with various parties involved are the basis for this chapter. Dr. Gardner read the manuscript of this chapter and gave me permission to use this information.

1. Curtis Tucker, "Okeene Rattlesnake Hunt," Enid Blog, last updated April 9, 2023, https://www.enidbuzz.com/okeene-rattlesnake-hunt/.
2. National Research Council (U.S.) and Institute of Medicine (U.S.) Committee on the Use of Laboratory Animals in Biomedical and Behavioral Research, *Use of Laboratory Animals in Biomedical and Behavioral Research* (Washington, DC: National Academies Press, 1988).
3. "US Animal Research Statistics," Speaking of Research, last updated August 23, 2021, https://speakingofresearch.com/facts/statistics/.
4. Philipp Schwedhelm et al., "How Many Animals Are Used for SARS-CoV-2 Research?: An Overview on Animal Experimentation in Pre-Clinical and Basic Research," EMBO Reports 22, no. 10 (October 2021): e53751, https://doi.org/10.15252%2Fembr.202153751.
5. Sung-Yoon Kang, Jeongmin Seo, and Hye-Ryun Kang, "Desensitization for the Prevention of Drug Hypersensitivity Reactions," *Korean Journal of Internal Medicine* 37, no. 2 (March 2022): 261–70, https://doi.org/10.3904%2Fkjim.2021.438.
6. Scott R. Loss, Tom Will, and Peter P. Marra, "The Impact of Free-Ranging Domestic Cats on Wildlife of the United States," *Nature Communications* 4 (2013): 1396, https://doi.org/10.1038/ncomms2380.
7. Scott R. Loss, Tom Will, and Peter P. Marra, "Estimation of Bird-Vehicle Collision Mortality on U.S. Roads," *Journal of Wildlife Management* 78, no. 5 (2014): 763–71, http://dx.doi.org/10.1002/jwmg.721; Mark Matthew Braunstein, "U.S. Roads Kill a Million a Day: Driving Animals to Their Graves," *Culture Change*, accessed August 30, 2023, http://www.culturechange.org/issue8/roadkill.htm; Wikipedia, s.v. "Roadkill," last updated August 15, 2023, https://en.wikipedia.org/wiki/Roadkill.
8. Wikipedia, s.v. "Animal slaughter," last updated August 30, 2023, https://en.wikipedia.org/wiki/Animal_slaughter.

9 Kent H. Redford, "The Empty Forest," *BioScience* 42, no. 6 (June 1992): 412–22, https://doi.org/10.2307/1311860.

10 Hannah Ritchie and Max Roser, "Fishing and Overfishing," Our World in Data, last updated October 1, 2021, https://ourworldindata.org/fish-and-overfishing.

11 "Threats to Pangolins," Wildlife Conservation Network, Pangolin Crisis Fund, last updated June 11, 2022, https://pangolincrisisfund.org/threats-to-pangolins/.

12 Curtis Tucker, "Okeene Rattlesnake Hunt," Enid Blog, last updated April 9, 2023, https://www.enidbuzz.com/okeene-rattlesnake-hunt/.

13 George R. Pisani and Barbara R. Stephenson, "Food Habits in Oklahoma of *Crotalus atrox* in Fall and Early Spring," *Transactions of the Kansas Academy of Science* 94, no. 3–4 (1991): 137–41, https://doi.org/10.2307/3627861.

14 Roberts, Janovy, and Nadler, *Foundations of Parasitology*, 536.

15 J. Teague Self and Robert E. Kuntz, "Host-Parasite Relations in Some Pentastomida," *Journal of Parasitology* 53, no. 1 (February1967): 202–6, https://doi.org/10.2307/3276647.

16 Gramlich, "What the Data Says."

17 John Janovy Jr., *Dunwoody Pond: Reflections on the High Plains Wetlands and the Cultivation of Naturalists* (New York: St. Martin's Press, 1994), ix–xvi.

Chapter Nine: Profiting from Parasites?

This chapter uses parasites of beetles, especially the single-celled ones known as gregarines, to illustrate something every biologist knows: non-human organisms can and do perform functions that humans cannot. That statement could probably be considered a general principle of biology. Most of this discussion is based on the research of Richard Clopton, a doctoral student in my lab in the early 1990s, and on the various conversations held Friday afternoons at Barry's Bar and Grill in Lincoln, Nebraska.

1 Marina A. Costa et al., "Intestinal Parasites in Paper Money Circulating in the City of Diamantina (Minas Gerais, Brazil)," *Research and Reports in Tropical Medicine* 9 (May 2018): 77–80, https://doi.org/10.2147/rrtm.s157896.

2 "The Nobel Prize in Physiology or Medicine 2015," Nobel Prize, accessed August 30, 2023, https://www.nobelprize.org/prizes/medicine/2015/summary/.

3 Alan Huffman, "The Ethics of Using Off-Label Medications to Treat COVID-19," *Annals of Emergency Medicine* 79, no. 6 (June 2022): A13–A15, https://doi.org/10.1016/j.annemergmed.2022.04.007.

4 "Moderna COVID-19 Vaccine Storage and Handling Summary," Centers for Disease Control and Prevention, U.S. Department of Health and Human Services, last updated March 30, 2023, https://www.cdc.gov/vaccines/covid-19/info-by-product/moderna/downloads/storage-summary.pdf.

5 Kelsey Vaughan et al., "The Costs of Delivering Vaccines in Low- and Middle-Income Countries: Findings from a Systematic Review," *Vaccine: X* 2 (July 2019): 100034, https://doi.org/10.1016/j.jvacx.2019.100034.

6 "Malaria," World Health Organization, last updated March 27, 2023, https://www.who.int/news-room/questions-and-answers/item/malaria.

7 Nadine Moeller, "Archaeologists Find Silos and Administration Center from Early Egyptian City," *University of Chicago News*, July 1, 2008, https://news.uchicago.edu/story/archaeologists-find-silos-and-administration-center-early-egyptian-city; Wikipedia, s.v. "Joseph's Granaries," last updated August 29, 2023, https://en.wikipedia.org/wiki/Joseph%27s_Granaries.

8 Muez Berhe et al., "Post-Harvest Insect Pests and Their Management Practices for Major Food and Export Crops in East Africa: An Ethiopian Case Study," *MDPI* 13, no. 11 (November 2022): 1068, https://doi.org/10.3390/insects13111068.

9 John Steinbeck and Edward F. Ricketts, *Sea of Cortez: A Leisurely Journey of Travel and Research, with a Scientific Appendix Comprising Materials for a Source Book on the Marine Animals of the Panamic Faunal Provinces* (New York: Viking, 1941), 33.

10 Alina Chan and Matt Ridley, *Viral: The Search for the Origin of COVID-19* (New York: Harper Collins, 2001).

11 Roberts, Janovy, and Nadler, *Foundations of Parasitology*, 132–37.

12 Wikipedia, s.v. "Shmoo," last updated August 19, 2023, https://en.wikipedia.org/wiki/Shmoo.

13 Roberts, Janovy, and Nadler, *Foundations of Parasitology*, 4–6.

14 Clopton, Percival, and Janovy, "*Gregarina niphandrodes*," 472–79.

15 Richard E. Clopton, "Specificity in the Gregarine Assemblage Parasitizing *Tenebrio molitor*" (PhD dissertation, University of Nebraska–Lincoln, 1993).

16 Roberts, Janovy, and Nadler, *Foundations of Parasitology*, 411–16.

17 Long Zhang and Michel Lecoq, "*Nosema locustae* (Protozoa, Microsporidia), a Biological Agent for Locust and Grasshopper Control," *Agronomy* 11, no. 4 (2021): 711–23, https://doi.org/10.3390/agronomy11040711.

18 Carlos E. Lange, "Usefulness of Protozoa for the Biological Control of Acridians (Orthoptera: Acridoidea)," *Revista de la Sociedad Entomologica Argentina* 58, no. 1–2 (1999): 26–33, https://www.biotaxa.org/RSEA/article/view/32657; Jeffrey A. Lockwood, Charles R. Bomar, and Al B. Ewen, "The History of Biological Control with *Nosema locustae*: Lessons for Locust Management," *Insect Science and Its Application*

19, no. 4 (1999): 333–50, https://doi.org/10.1017/S1742758400018968.

19 Roberts, Janovy, and Nadler, *Foundations of Parasitology*, 129–31.

Chapter Ten: Multiple-Kind Lotteries

The student whose story opens this chapter is actually a composite of several who worked on parasites of small rodents from the late 1970s through the 1980s, but her described experience is exactly representative of the others. The first of these studies was done by a student named Gail Beaty, who I was able to contact and have a conversation about her project done so many years ago. Her data analysis inspired us to develop software to analyze parasite communities much larger than those of mice. The programming contest was with Richard Clopton when he was a doctoral student in my lab; his brother David, an electrical engineer, eventually solved the programming problem. At the time these students were doing their work, the parasitological literature included extensive discussion of whether the multiple species living in and on a single host were a community, in the sense that these species interacted, or an assemblage and thus more or less independent of one another. This discussion drove the effort to bring that software into publication. The "Three Blind Mice" nursery rhyme evidently originated in the early seventeenth century as a singing exercise and may have been written by Thomas Ravenscroft. See the Wikipedia link below.

1 Wikipedia, s.v. "Three Blind Mice," last updated July 19, 2023, https://en.wikipedia.org/wiki/Three_Blind_Mice.

2 Hermann Hesse, *Das Glasperlenspeil* (Zurich: Fretz and Wasmuth Verlag, 1943; translated by Mervyn Savill in 1949 and published as *The Glass Bead Game*, New York: Holt, Reinhart and Winston, 1969).

3 John Janovy Jr. et al., "Species Density Distributions as Null Models for Ecologically Significant Interactions of Parasite Species in an Assemblage," *Ecological Modeling* 77, no. 2–3 (February 1995):

189–96, https://doi.org/10.1016/0304-3800(93)E0087-J; John Janovy Jr., "Defining the Field: Concurrent Infections and the Community Ecology of Helminth Parasites," *Journal of Parasitology* 88, no. 3 (June 2002): 440–45, https://doi.org/10.2307/3285429.

4 John Janovy Jr., *Bernice and John: Finally Meeting Your Parents Who Died a Long Time Ago* (pdf e-book, Amazon Kindle edition, 2014), chap. 1.

5 Janovy, "Defining the Field," 440–45.

6 Hermann Hesse, *The Glass Bead Game,* trans. Mervyn Savill (New York: Holt, Reinhart and Winston, 1969).

7 Janovy et al., "Species Density Distributions," 189–96.

8 "Daily Mobile SMS Usage Statistics," SMSEagle, March 6, 2017, https://www.smseagle.eu/2017/03/06/daily-sms-mobile-statistics/; Petroc Taylor, "Total Number of SMS and MMS Messages Sent in the United States from 2005 to 2021," Statista, last updated January 18, 2023, https://www.statista.com/statistics/185879/number-of-text-messages-in-the-united-states-since-2005/.

Chapter Eleven: Iron Wheels

Matthew G. Bolek, the subject of this chapter, is a faculty member at Oklahoma State University where he continues to demonstrate his creativity in research on topics some people might consider arcane beyond description but that in fact teach us much about the fundamental nature of parasitism. Bolek is winner of the American Society of Parasitologists' H. B. Ward Medal, the highest honor ASP can give to a member based on research, and he has also served as president of that society. The ascaris life cycle diagrams are based on those from *Foundations of Parasitology*, ninth edition, originally drawn by William Ober and Claire Garrison. In my view, chapter 11 is probably the best explanation of why we study parasitism, aside from the fact that parasites are so inherently fascinating in their own right.

1 Email from Joe Pollock, sent to John Janovy Jr., on Wednesday, January 2, 2015.
2 Roberts, Janovy, and Nadler, *Foundations of Parasitology*.
3 Matthew G. Bolek, Heather A. Stigge, and Kyle D. Gustafson, "The Iron Wheel of Parasite Life Cycles: Then and Now!," in *A Century of Parasitology: Discoveries, Ideas and Lessons Learned by Scientists who Published in* The Journal of Parasitology, *1914–2014*, eds. John Janovy Jr. and Gerald W. Esch (Chichester, UK: John Wiley and Sons, 2016), 131–47.
4 Matthew G. Bolek and John Janovy Jr., "*Rana catesbeiana* (Bullfrog) Gigantic Tadpole," *Herpetological Review* 35 (2004): 376–77.
5 Matthew G. Bolek and John Janovy Jr., "Evolutionary Avenues for, and Constraints on, the Transmission of Frog Lung Flukes (*Haematoloechus* spp.) in Dragonfly Second Intermediate Hosts," *Journal of Parasitology* 93, no. 3 (June 2007): 593–607, https://doi.org/10.1645/ge-1011r.1.
6 Karen Armstrong, *A Short History of Myth* (Edinburgh: Canongate Books, 2005), 124–30.
7 Oliver Wilford Olson, *Animal Parasites: Their Life Cycles and Ecology* (New York, Dover, 1974), 292–96.
8 Scott D. Snyder and John Janovy Jr., "Second Intermediate Host Specificity of *Haematoloechus complexus* and *Haematoloechus medioplexus* (Digenea: Haematoloechidae)," *Journal of Parasitology* 80, no. 6 (December 1994): 1052–55, https://doi.org/10.2307/3283461; Scott D. Snyder and John Janovy Jr., "Behavioral Basis of Second Intermediate Host Specificity among Four Species of *Haematoloechus* (Digenea: Haematoloechidae)," *Journal of Parasitology* 82, no. 1 (February 1996): 94–99, https://doi.org/10.2307/3284122.

9 Bolek and Janovy, “Evolutionary Avenues,” 593–607.

10 Leonard Knollenberg and Stephan Sommer, “Diverging Beliefs on Climate Change and Climate Policy: The Role of Political Orientation,” *Environmental and Resource Economics* 84, no. 4 (2023): 1031–49, https://doi.org/10.1007/s10640-022-00747-1.

11 Douglas Hofstadter, *Metamagical Themas: Questing for the Essence of Mind and Pattern* (New York: Basic Books, 1985).

Chapter Twelve: Jigsaw Puzzles

The inspiration for this chapter was the Gerald D. Schmidt Memorial Lecture delivered by Richard Clopton in September 2022 at the Cedar Point Biological Station in Keith County, Nebraska, during the annual meeting of the Rocky Mountain Conference of Parasitologists. The jigsaw puzzle figure and the way Dr. Clopton used it were more effective at telling his audience why we will forever remain ignorant about Earth and its inhabitants than any other presentation I'd seen. Dr. Janine Caira, University of Connecticut, whose research on elasmobranch tapeworms is a remarkable lifetime achievement, also read a draft of this chapter.

1 Wikipedia, s.v. “There are Unknown Unknowns,” last updated January 5, 2024, https://en.wikipedia.org/wiki/There_are_unknown_unknowns.

2 Wikipedia, s.v. “*The Theory of Island Biogeography*,” last updated July 5, 2023, https://en.wikipedia.org/wiki/The_Theory_of_Island_Biogeography.

3 Roberts, Janovy, and Nadler, *Foundations of Parasitology*, 9–21.

4 Roberts, Janovy, and Nadler, *Foundations of Parasitology*, 1–8.

5 About the “Rocky Mountain Conference of Parasitologists,” Rocky Mountain Conference of Parasitologists, accessed August 30, 2023, http://sites.coloradocollege.edu/rmcp/.

6 Sarah Richardson and John Janovy Jr., "*Actinocephalus carrilynnae* n. sp. (Apicomplexa: Eugregarinorida) from the Blue Damselfly, *Enallagma civile* (Hagen)," *Journal of Protozoology* 37, no. 6 (November 1990): 567–70, https://doi.org/10.1111/j.1550-7408.1990.tb01266.x.

7 "Threats to the Amazon," Amazon Conservation Association, last updated April 30, 2020, https://www.amazonconservation.org/the-challenge/threats/.

8 "True Flies (Diptera)," Smithsonian Institution, accessed August 30, 2023, https://www.si.edu/spotlight/buginfo/true-flies-diptera.

9 Fan Zhang et al., "*Caenorhabditis elegans* as a Model for Microbiome Research," *Frontiers in Microbiology* 8 (2017): 1–10, https://doi.org/10.3389/fmicb.2017.00485.

10 Roberts, Janovy, and Nadler, *Foundations of Parasitology*, 377–463.

11 Michele L. Lemons, "An Inquiry-Based Approach to Study the Synapse: Student-Driven Experiments Using *C. elegans*," *Journal of Undergraduate Neuroscience Education* 15, no. 1 (Fall 2016): A44–A55, https://www.ncbi.nlm.nih.gov/pmc/articles/PMC5105963/.

12 Karin Kiontke and David H. A. Fitch, "Nematodes," *Current Biology* 23, no. 19 (2013): R862–R864, https://doi.org/10.1016/j.cub.2013.08.009.

13 Guinevere O. Drabik and Scott L. Gardner, "A New Species of *Ancylostoma* (Nemata: Strongylida: Ancylostomatidae) from Two Species of *Ctenomys* in Lowland Bolivia," *Journal of Parasitology* 105, no. 6 (December 2019): 904–12, http://dx.doi.org/10.1645/19-100.

14 "BIOSIS Citation Index," via University of Nebraska–Lincoln Libraries, accessed August 29, 2023, https://clarivate.com/webofsciencegroup/solutions/webodscience-biosis-citation-index/.

15 "BIOSIS Citation Index."

16 Richard E. Clopton, Tamara J. Percival, and John Janovy Jr., "*Nubenocephalus nebraskensis* n. gen., n. sp. (Apicomplexa:

Actinocephalidae) from Adults of *Argia bipunctulata* (Odonata: Zygoptera)," *Journal of Parasitology* 79, no. 4 (August 1993): 533–37, https://doi.org/10.2307/3283378; Richard E. Clopton, Tamara J. Percival, and Jerry L. Cook, "*Naiadocystis phykoterion* n. gen., n. sp. (Apicomplexa: Eugregarinida: Hirmocystidae), from the Mexican Pygmy Grasshopper, *Paratettix mexicanus* (Orthoptera: Tetrigidae), in the Texas Big Thicket with Recognition of Three Previously Described Species of *Naiadocystis*," *Journal of Parasitology* 90, no. 2 (April 2004): 301–7, https://doi.org/10.1645/ge-137r.

17 Janine N. Caira and Kirsten Jensen, "Diversity and Phylogenetic Relationships of 'tetraphyllidean' Clade 3 (Cestoda) based on New Material from Orectdolobiform Sharks in Australia and Taiwan," *Folia Parasitologica* 69 (2022): 010, https://doi.org/10.14411/fp.2022.010.

18 Caira and Jensen, "Diversity and Phylogenetic Relationships."

19 Paul R. Ehrlich, *The Population Bomb* (New York: Ballantine Books, 1968), 24–29; "Dr. Paul R Ehrlich on 60 Minutes CBS," Climate Alert, January 1, 2023, YouTube video, 4:59, https://www.youtube.com/watch?v=l_j2QVg6e00.

Chapter Thirteen: Manipulation

My sincere thanks to Dr. Janice Moore, who joined Kaylen Michaelis and me for a long Zoom conversation about her work on parasite manipulation of host behavior. Dr. Moore looked over a draft of this chapter and offered some suggestions, which I incorporated.

1 Jacob Abbott, *Gentle Measures in the Management and Training of the Young, or, The Principles on Which a Firm Parental Authority May Be Established and Maintained, without Violence or Anger, and the Right Development of the Moral and Mental Capacities Be Promoted by*

Methods in Harmony with the Structure and the Characteristics of the Juvenile Mind (Harper and Brothers, 1871).

2 Lora Rickard Ballweber, "*Dicrocoelium dendriticum* in Ruminants (Lancet Fluke, Lesser Liver Fluke)," Merck Manual/Veterinary Manual, last updated October 1, 2022, https://www.merckvetmanual.com/digestive-system/fluke-infections-in-ruminants/dicrocoelium-dendriticum-in-ruminants.

3 Shiraz Tyebji et al., "Impaired Social Behavior and Molecular Mediators of Associated Neural Circuits during Chronic *Toxoplasma gondii* Infection in Female Mice," *Brain Behavior and Immunity* 80 (August 2019): 88–108, https://doi.org/10.1016/j.bbi.2019.02.028.

4 "Social Media and the Spread of Harmful Ideas," Cornell University Course blog for INFO 2040/CS 2850/Econ 2040/SOC 2090, last updated November 27, 2017, https://blogs.cornell.edu/info2040/2017/11/27/social-media-and-the-spread-of-harmful-ideas/.

5 Ballweber, "*Dicrocoelium dendriticum*."

6 Oriel FeldmanHall et al., "Empathic Concern Drives Costly Altruism," *NeuroImage* 105 (January 2015): 347–56, https://doi.org/10.1016/j.neuroimage.2014.10.043.

7 Shailesh Shrestha et al., "Financial Impacts of Liver Fluke on Livestock Farms under Climate Change—A Farm Level Assessment," *Frontiers in Veterinary Science* 7 (2020): 564795, https://doi.org/10.3389/fvets.2020.564795.

8 William M. Bethel and John C. Holmes, "Altered Evasive Behavior and Responses to Light in Amphipods Harboring Acanthocephalan Cystacanths," *Journal of Parasitology* 59, no. 6 (December 1973): 945–56, https://doi.org/10.2307/3278623.

9 Bethel and Holmes, "Altered Evasive Behavior," 945–56.

10 William M. Bethel and John C. Holmes, "Correlation of Altered Evasive

Behavior in Gammarus lacustris (Amphipoda) Harboring Cystacanths of Polymorphus paradoxus (Acanthocephala) with Infectivity to the Definitive Host," *Journal of Parasitology* 60, no. 2 (April 1974): 272-274.

11 Moises Velasquez-Manoff, "The Anti-Vaccine Movement's New Frontier," *New York Times Magazine*, May 5, 2022, https://www.nytimes.com/2022/05/25/magazine/anti-vaccine-movement.html; "History of Anti-Vaccination Movements," College of Physicians of Philadelphia/Mütter Museum, last updated April 20, 2022, https://historyofvaccines.org/vaccines-101/misconceptions-about-vaccines/history-anti-vaccination-movements.

12 Vanessa O. Ezenwa et al., "Host Behaviour—Parasite Feedback: An Essential Link Between Animal Behaviour and Disease Ecology," *Proceedings of the Royal Society B* 283, no. 1828 (April 2015): 20153078, http://doi.org/10.1098/rspb.2015.3078.

13 Janice Moore, "Responses of an Avian Predator and Its Isopod Prey to an Acanthocephalan Parasite," *Ecology* 64, no. 5 (October 1983): 1000–1015, https://doi.org/10.2307/1937807.

14 Moore, "Responses of an Avian Predator," 1000–1015.

15 Wikipedia, s.v. "Eugene Schieffelin," last updated August 5, 2023, https://en.wikipedia.org/wiki/Eugene_Schieffelin.

16 "European Starling," New York Invasive Species Information, accessed August 30, 2023, https://nyis.info/invasive_species/european-starling.

17 Moore, "Responses of an Avian Predator," 1000–1015.

18 Moore, "Responses of an Avian Predator," 1000–1015.

19 Wikipedia, s.v. "Attempts to Overturn the 2020 United States Presidential Election," last updated August 30, 2023, https://en.wikipedia.org/wiki/Attempts_to_overturn_the_2020_United_States_presidential_election; Philip Bump, "The 2020 Election Was Neither Stolen Nor Rigged: A Primer," *Washington*

Post, September 15, 2022, https://www.washingtonpost.com/politics/2022/09/15/2020-election-trump-false-fraud-claims/.

20 "CDC Museum COVID-19 Timeline," Centers for Disease Control and Prevention, U.S. Department of Health and Human Services, last updated March 15, 2023, https://www.cdc.gov/museum/timeline/covid19.html.

21 Chris Jones, "What Really Happened at the Capitol on January 6?," Belt Magazine, January 20, 2021, https://beltmag.com/capitol-january-6-insurrection/; Andrea Raballo, Michele Poletti, and Antonio Preti, "Vaccine Hesitancy, Anti-Vax, COVID-Conspirationism: From Subcultural Convergence to Public Health and Bioethical Problems," *Frontiers in Public Health*, May 9, 2022, https://www.frontiersin.org/articles/10.3389/fpubh.2022.877490/full.

22 Nico Martinez, "NBA Legend John Stockton Claims 'Thousands' of Athletes Have Died from the Vaccine," Fadeaway World, accessed August 30, 2023, https://www.msn.com/en-us/sports/more-sports/nba-legend-john-stockton-claims-thousands-of-athletes-have-died-from-the-vaccine/ar-AA15NoAh?ocid=mailsignout&pc=U591&cvid=4ef04fa2370c468981ffd0ed9ec412c1; Monica Pivetti, Giannino Melotti, and Claudia Mancini, "Vaccines and Autism: A Preliminary Qualitative Study on the Beliefs of Concerned Mothers in Italy," *International Journal of Qualitative Studies on Health and Well-being* 15, no. 1 (2020): 1754086–101, https://doi.org/10.1080/17482631.2020.1754086.

23 Glen Harlan Reynolds, "Social Media Threat: People Learned to Survive Disease: We Can Handle Twitter," *USA Today*, last updated November 24, 2017, https://www.usatoday.com/story/opinion/2017/11/20/social-media-threat-people-survived-disease-we-can-handle-twitter-glenn-reynolds-column/879185001/.

24 Luigi L. Cavalli-Sforza and Marcus W. Feldman, *Cultural Transmission and Evolution: A Quantitative Approach* (Princeton, NJ: Princeton University Press, 1981), 31–76.

25 Hofstadter, *Metamagical Themas*, 49–67.

26 Wikipedia, s.v. "Acquisition of Twitter by Elon Musk," last updated August 27, 2023, https://en.wikipedia.org/wiki/Acquisition_of_Twitter_by_Elon_Musk; David Folkenflik and Mary Yang, "Fox News Settles Blockbuster Defamation Lawsuit with Dominion Voting Systems," Nebraska Public Media, April 18, 2023, https://www.npr.org/2023/04/18/1170339114/fox-news-settles-blockbuster-defamation-lawsuit-with-dominion-voting-systems.

27 Douglas Hofstadter and Emmanuel Sander, *Surfaces and Essences: Analogy as the Fuel and Fire of Thinking* (New York: Basic Books, 2013), 257–315.

28 "Teen Watched Simulated Hanging Video on Instagram before Suicide," CBS News/60 Minutes Overtime, December 11, 2022, https://www.cbsnews.com/news/instagram-hanging-video-suicide-60-minutes-2022-12-11/; "Helping Your Teen Navigate Instagram Safely," Parental Guide for Teens on Instagram, accessed August 30, 2023, https://about.instagram.com/community/parents.

29 Sunny Tsai, "Gov. Abbott Bans TikTok across State Agency Devices," Spectrum News, December 8, 2022, https://spectrumlocalnews.com/tx/south-texas-el-paso/news/2022/12/08/gov--abbott-bans-tiktok-across-state-agency-devices.

Chapter Fourteen: General Theory of Infectivity

This chapter should be considered a logical consequence of reading the books by Hofstadter, Farb, and Cavalli-Sforza and Feldman referenced above and below. I read those books when they first came out in the 1980s, acquired copies for my personal library, and have been thinking about their implications ever since. This chapter is also fairly representative of the kinds of conversations taking place in my laboratory and at social occasions outside work since the mid-1960s.

1 Response generated by ChatGPT, OpenAI, March 7, 2023, https://chat.openai.com/chat.

2 Bryce Ryan and Neal C. Gross, "The Diffusion of Hybrid Seed Corn in Two Iowa Communities," *Rural Sociology* 8, no. 1 (1943): 15–24.

3 Rokon Zaman, "Innovation S-curve—Episodic Evolution," The Waves: Technology, Society, and Policy, last updated March 13, 2022, https://www.the-waves.org/2022/03/13/innovation-s-curve-episodic-innovation-evolution/.

4 Peter Farb, *Word Play: What Happens When People Talk* (New York: Alfred A. Knopf, 1973), 38–57.

5 Wikipedia, s.v. "Compartmental models in epidemiology," last updated August 24, 2023, https://en.wikipedia.org/wiki/Compartmental_models_in_epidemiology.

6 Samuel Mwalili et al., "SEIR Model for COVID-19 Dynamics Incorporating the Environment and Social Distancing," *BMC Research Notes* 13, no. 1 (2020): 352, https://doi.org/10.1186/s13104-020-05192-1; Alberto Godio, Francesca Pace, and Andrea Vergnano, "SEIR Modeling of the Italian Epidemic of SARS-CoV-2 Using Computational Swarm Intelligence," *International Journal of Environmental Research and Public Health* 17, no. 10 (2020): 3535, https://doi.org/10.3390/ijerph17103535.

7 Richard Dawkins, *The Selfish Gene*, 40th anniversary ed. (Oxford, UK: Oxford University Press, 2016), 245–60.

8 Farb, *Word Play*, 38–57.

9 Mark Abadi, "Trump Has Used Some Bizarre Words and Phrases

That Left People Scratching Their Heads—Here Are 8 of the Worst," Business Insider, November 30, 2017, https://www.businessinsider.com/trump-made-up-words-confusing-phrases-2017-11#covfefe-2; Hannah Natanson, John Woodrow Cox, and Perry Stein, "Trump's Words, Bullied Kids, Scarred Schools," *Washington Post*, February 13, 2020, https://www.washingtonpost.com/graphics/2020/local/school-bullying-trump-words/.

10 Roberts, Janovy, and Nadler, *Foundations of Parasitology*, 625.

11 "American Society of Parasitologists," American Society of Parasitologists, accessed August 30, 2023, https://www.amsocparasit.org/.

12 Andy Slavitt, *Preventable: The Inside Story of How Leadership Failures, Politics, and Selfishness Doomed the U.S. Coronavirus Response* (New York: St. Martin's Press, 2021), 88–94; "Timeline of Trump's Coronavirus Responses," Lloyd Doggett, U.S. Representative, March 2, 2022, https://doggett.house.gov/media/blog-post/timeline-trumps-coronavirus-responses.

INDEX

A

D

E

F

G

H

I

L

M

N

O

R

S

T

U

V

W

ABOUT THE AUTHOR

John Janovy Jr. is a successful author, respected scientist, artist, and award-winning teacher. He retired from the University of Nebraska–Lincoln, where he was Varner Distinguished Professor of Biological Sciences. His twenty books include those with a natural history theme, reflections on high school athletics, anti-intellectualism in America, higher education, and travel. He is the coauthor of *Foundations of Parasitology*, the leading textbook in his discipline, and senior editor of *A Century of Parasitology: Discoveries, Ideas and Lessons Learned by Scientists who Published in* The Journal of Parasitology, *1914–2014*. Janovy's honors include the University of Nebraska Distinguished Teaching Award, UNL's Outstanding Research and Creativity Award, and the American Society of Parasitologists Clark P. Read Mentorship Award. Janovy served as director of the Cedar Point Biological Station and as interim director of the University of Nebraska State Museum. His teaching experiences include service in large-enrollment introductory courses in addition to upper division and graduate seminars. He supervised eighteen MS and fourteen PhD students and many undergraduate researchers, including two Fulbright Scholarship winners. He and his students have published over a hundred scientific papers. His website is https://www.johnjanovy.com